An Interactive History of the Clean Air Act

An Interactive History of the Clean Air Act
Scientific and Policy Perspectives

Jonathan Davidson
Environmental Research Institute
University of California, Riverside, CA
USA

Joseph M. Norbeck
Bourns College of Engineering
University of California, Riverside, CA
USA

AMSTERDAM • BOSTON • HEIDELBERG • LONDON • NEW YORK • OXFORD
PARIS • SAN DIEGO • SAN FRANCISCO • SINGAPORE • SYDNEY • TOKYO

Elsevier
32 Jamestown Road, London NW1 7BY
225 Wyman Street, Waltham, MA 02451, USA

First edition 2012

Notices
Knowledge and best practice in this field are constantly changing. As new research and experience broaden our understanding, changes in research methods, professional practices, or medical treatment may become necessary.

Practitioners and researchers must always rely on their own experience and knowledge in evaluating and using any information, methods, compounds, or experiments described herein. In using such information or methods they should be mindful of their own safety and the safety of others, including parties for whom they have a professional responsibility.

To the fullest extent of the law, neither the Publisher nor the authors, contributors, or editors, assume any liability for any injury and/or damage to persons or property as a matter of products liability, negligence or otherwise, or from any use or operation of any methods, products, instructions, or ideas contained in the material herein.

British Library Cataloguing-in-Publication Data
A catalogue record for this book is available from the British Library

Library of Congress Cataloging-in-Publication Data
A catalog record for this book is available from the Library of Congress

ISBN: 978-0-323-16541-9

This book has been manufactured using Print On Demand technology. Each copy is produced to order and is limited to black ink. The online version of this book will show color figures where appropriate.

Contents

Preface

The Clean Air Act amendments of 1970 fundamentally reframed regulatory relationships among government and private sector stakeholders. This impressive legislative collaboration mandated national quality standards, certification of state implementation plans, and technological innovation by affected industries. An emergent Environmental Protection Agency (EPA) developed health-based standards based on controlled epidemiologic studies. States responded by dramatically expanding their air quality regulatory programs. Major corporations escalated environmental issues to top-level management. Their trade associations advocated common concerns before governmental bodies while conveying policy developments to their constituents.

One of us (Norbeck) started working at Ford Motor Company's Scientific Research Laboratory in the early 1970s and recognized the change in the automotive and petroleum industries attitude toward environmental issues as one from pejorative to cooperative in working toward meaningful solutions. These regulations affected changes in not only industry but state and federal agencies as well. He joined the University of California Riverside in 1992 to become the founding director of UCR's College of Engineering Center for Environmental Research and Technology and later the Environmental Research Institute. The idea of documenting the change in industry and government agencies spawned from these experiences.

Since 1973, coauthor Davidson's professional involvement includes roles as a planner, attorney, instructor, and mediator on environmental matters. His experiences as an advocate before and within governments, and in facilitating multiparty negotiations, inform the interactive models that frame the book's overall structure.

Underlying research support was provided by a grant from the Andrew J. Mellon Foundation. Environmental Research Institute staff surveyed officials at EPA, state air quality programs, and selected corporations from primary impacted sectors (automotive, chemical, aerospace, and consumer goods). The chapters that follow incorporate these data within the broader contexts of stakeholder interactions in environmental policy development and how scientific analyses impact these decisions.

The authors wish to thank Georgia Elliott for her support of this project and in helping acquire the funding from the Mellon Foundation to initiate the program. Regina Hazlinger supervised the numerous students that documented the historical information from the state and local regulatory agencies and industries.

The manuscript that follows tracks stakeholder interactions through four decades of Clean Air Act implementation. Its publication at a time when partisanship pervades domestic politics stands in sharp contrast to the collaborative optimism of the 1970 law. The authors hope that readers may gain insights by referencing its findings to contexts where multiple stakeholders confront intertwined science-policy issues.

Jonathan Davidson
Joseph M. Norbeck
October 2011

1 Introduction and Overview

The United States Clean Air Act of 1970 (Clean Air Act) raised unprecedented implementation challenges for governments and affected industries. Nationally, the Environmental Protection Agency (EPA) established health-based quality standards and directed states to implement control plans subject to federal certification. States responded with program initiatives to attain these levels within designated time frames. Companies initiated emission control technologies and production processes by necessity to comply with evolving federal and state regulations. Many escalated environmental issues to top-level management concerns. Automotive, chemical, and other industry-wide associations emerged as intermediaries on policy and technical issues. Comparably, states formed regional coalitions to address cross-jurisdictional air quality impacts.

This book traces the adaptive administrative and technological responses leading to the current framework for air pollution control policy. Through four decades, public–private and intergovernmental relationships continually center around three primary issue areas:

- the extent to which health-based scientific studies provide a basis for emission controls [National Ambient Air Quality Standards (NAAQS) and regulatory standards for toxic pollutants],
- the extent to which states must plan for and regulate air emissions subject to federal certification and sanctioning authority, and
- the extent to which national and state policies can require industries to implement timely emission control measures and technologies to meet compliance standards.

These concerns provide the underpinnings for ongoing Clean Air Act implementation.

Concurrently, the dynamics among EPA, states, industries, trade associations, and other stakeholders have evolved toward increased reliance on collaborative processes as first-choice arenas. These alternative contexts encourage open dialogue on scientific findings, government-implementation roles, and for determining acceptable control technologies.

1.1 The Historical Context for Clean Air Act Implementation

Until the mid-1960s, the federal role in air quality management focused on research and technical assistance. Pollution control administration remained in the domain of state and local administrators. While industries increased their attention to environmental concerns, these programs had limited authority within most management structures.

An Interactive History of the Clean Air Act. DOI: 10.1016/B978-0-12-416035-4.00001-1

The 1970 Clean Air Act emerged as a remarkable bipartisan consensus among Congressional leaders. Its amendment of earlier legislation melded aspirations to protect public health and welfare, to implement national air quality standards, and to trust that American industrial innovation could reduce emission of harmful pollutants. An environmental initiative of this magnitude was exceptional at the time, and perhaps unimaginable in present political contexts.

Through the 1970s, litigation and intense Congressional debate became the primary forums for defining the Act's overall implementation structure. Supreme Court decisions sustained EPA authority to issue national air quality standards. The Court also affirmed that states could not allow economic or technological limitations to trump regulatory measures. Further, states would be barred from allowing "significant deterioration" in areas already compliant with national standards. Amendments in 1977 codified this policy and extended compliance deadlines for automobile manufacturers and power plants.

Retrenchment in administrative commitment to environmental issues was an overriding marker in the 1980s.The EPA cut its budget, staff, and assistance to states as part of broader directives to reduce federal oversight. However, there were quieter innovations in state program approaches and cooperative agreements among automobile manufacturers to exchange pre-competitive research on emission reduction technologies.

Clean Air Act amendments in 1990 directed the EPA to regulate 189 toxic pollutants at their source. The scope of these provisions extended beyond larger-scale industries to businesses such as dry cleaners and photocopy stores. Pursuant to Title V, the agency developed performance standards for pollutants specified within this law through negotiations with trade associations, impacted businesses, and other stakeholders. This program also directed states to administer and finance these regulations.

The EPA's addition of more stringent particulate standards in 1997 created widespread impacts on transportation planning as well as on industries and businesses. In December 2009, the agency designated greenhouse gases (CO_2, methane, nitrous dioxide, and chlorofluorohydrocarbons; GHG) as pollutants within the scope of Clean Air Act standards. This controversial process required an initial threshold acceptance of scientific consensus that global climate change was accelerating. A second, and far more controversial decision, was that societal actions were causative factors in these changes. The next determination was to identify key sources of GHG emissions based on empirical data. From that point, policies at all governmental levels must identify appropriate actions. Implementation options may include regulation, taxation, incentives, and other measures. International agreements and industry-accepted standards are additional considerations.

1.2 An Adaptive Regulatory Framework

The Clean Air Act, as amended in 1970 and 1990, continues to reframe relationships among governments and industries. It assigns a federal role to develop ambient air quality standards with authority to guide, certify, and sanction state implementation programs. It also delegates authority to the EPA to direct an expansive regulatory program for toxic pollutants. States are charged with the bulk of administering these requirements. Industries and businesses must continually adapt and develop technologies to comply with state and federal permit requirements.

Relations among environmental stakeholders have evolved from openly antagonistic stances toward emphasis on collaborative interactions. In the 1970s, the EPA's authority for criteria pollutants, for protecting areas exceeding ambient air quality standards, and state implementation plan (SIP) guidance were resolved primarily via judicial challenges. The 1980s brought more collaborative exchange within industrial sectors. The Cooperative Research Act of 1984 provided an underlying framework for corporations to support pre-competitive scientific studies and develop pollution control technologies.

Following passage of the 1990 amendments, the EPA and states have adapted their strategies to promote more direct stakeholder involvement. Industrial and multi-state associations participate with environmental agency staffs through advisory and working groups. Formalized processes such as negotiated rulemaking provide alternatives to internal agency policy development prior to stakeholder participation. Federal–state, EPA–industry, and state–industry collaborative programs address shared air quality concerns.

Current relationships reflect the complex evolution of these interactions. The EPA's headquarters office determines standards for criteria pollutants and rules for SIPs. While retaining ultimate authority, the central office delegates many significant interactions with state programs to its 10 regional offices. National staff works more directly with industry associations, state and multi-state representatives, and non-governmental organizations on policy matters.

As currently structured, state environmental agencies retain central roles in planning and implementing air quality strategies. They have primary responsibilities for regulating emissions to meet their implementation plan goals. States may develop stricter standards provided that they can be supported by health impact research. For example, California developed motor vehicle emission standards based on its own findings. Each state must also administer and financially support the national Title V permit program for toxic emissions specified in the 1990 amendments. Multi-state consortiums convey broader regional concerns and participate in developing national policies.

Industries and utilities interact directly with states on regulatory matters. Trade associations convey their constituents' interests in direct interactions with EPA's national and regional staff. They also assist members in interpreting technologies required for regulatory compliance.

1.3 EPA and State Responses to Clean Air Act Mandates

Clean Air Act implementation led to dramatic institutional shifts for governments and affected industries. Initially, the EPA focused on setting standards for air quality and control technologies, and on guidance for state implementation planning. Since 1990, the agency's scope has expanded to incorporate a detailed regulatory structure for toxic air pollutants. The EPA's growth is reflected in a comparison between its first-year $1 billion budget and 4,000 employees[1] with a nearly $14 billion discretionary budget[2] for more than 17,000 employees in 2010.[3]

State programs have comparably expanded their primary implementation roles to meet NAAQS. SIPs are revised continually to ensure ongoing abatement in nonattainment areas and sustaining air quality in areas that meet current criteria. Multistate organizations such as the Northeast States for Coordinated Air Use Management and the Great Lakes Commission address common regional issues in controlling air emissions.

1.4 Industry and Trade Associations Respond to Technological Challenges

Industries have responded to Clean Air Act mandates with new technologies and escalated management attention. Beginning in the 1980s, motor vehicle and other manufacturers established collaborative research programs to improve overall efficiencies. Companies that responded proactively were able to market control technologies to others within their sectors. The overlay of federal Title V permits has required technologies and practices prescribed for comparable toxic source emitters.

As regulatory challenges expanded, trade associations emerged as intermediaries between members and government policy makers. Automobile, chemical, and other manufacturing interests interpret developments and advocate common interests. Corporate members balance the tensions of responding to industry-wide regulations while protecting their proprietary interests. Smaller-scale businesses such as dry cleaners, print shops, and paint retailers emulated this pattern to address technology standards for toxic pollutants. Trade associations have provided crucial assistance with common regulatory challenges, operating requirements, and applicable control technologies.

1.5 The Impact of Science–Policy Interactions on Clean Air Act Implementation

The Clean Air Act established two major programs with scientific and technological underpinnings. The 1970 amendments required that the NAAQS "accurately

[1] United States Environmental Protection Agency, "EPA's Budget and Spending," http://www.epa.gov/planandbudget/budget.html (accessed October 2011).

[2] EPA, *Fiscal Year 2010, Agency Financial Report*, Section 2, p. 9 (2011).

[3] Ibid., Section 1, p. 2.

reflect the latest scientific knowledge useful in indicating the kind and extent of all identifiable effects on public health or welfare."[4] The EPA responded by designating NAAQS for carbon monoxide, ozone, lead, nitrogen oxides, particulate matter, and for sulfur dioxide. These pollutants and their associated criteria were based on controlled studies measuring health impacts on at-risk persons. In developing health-based air quality standards, the agency accepted the underlying validity of its epidemiological studies.

In addition to the health-based quality standards for ambient air quality, the EPA has established rules designating acceptable technologies and practices for facilities that emit toxic air pollutants. These directives are formulated in conjunction with representatives for impacted source emitters. Typically, the EPA accepts consensus findings on technological feasibility and "best practices." These underlying processes demand scientific rigor and advancing technologies. In tracing the evolution of these policy and regulatory developments, the chapters that follow can be framed within the following considerations:

- Is there a causative relationship between a designated pollutant or pollutant class. The accounts for the GHG designation and human health impairment via air emissions?
- If a causative relationship is determined, what concentrations during a specified time period in ambient air are deemed sufficient to cause harm to human health?
- If NAAQS are established for designated pollutants, how will causation be attributed to identifiable mobile and stationary emission sources?
- Are there available or feasible technologies and/or practices that will lead to reduction of one or more specified pollutants?

In effect, negotiations over designating NAAQS and acceptable technologies for abating toxic pollutants are bounded by consensus acceptance of underlying science-based findings. The factors referenced above apply as well to correlating GHG emissions with accelerating climate change and to identifying sources subject to developing policies or regulations.

While the elements of scientific causation and public policy for global climate change are beyond the scope of this book, its interactive framework may provide a useful context for evaluating policy options. The following chapters review Clean Air Act implementation from the perspectives of its primary participating actors. Chapter 2 presents structural frameworks for evolving stakeholder interactions. Chapters 3 through 5 provide overviews of EPA, state, and selected industry responses to Clean Air Act mandates. Chapter 6 addresses the role of trade associations and multi-state coalitions in defining and implementing air pollution control policy. The concluding chapter offers observations on overall dynamics in four decades of Clean Air Act implementation. These findings cover both structural interactions and how science and technology are applied in these contexts. It is hoped that

[4] Clean Air Act 1970 amendments, 42 United States Code Section 4208.

any "lessons learned," or "not yet learned" can provide insights for future U.S. policy and for other nations confronting the challenges of environmentally sustainable development.[5]

[5] See National Academy of Sciences, Committee on Air Quality Management in the United States, *Air Quality Management in the United States*, at p. 26 (National Academy of Sciences, 2004). This committee was comprised of members of the Board on Environmental Studies and Toxicology, the Board on Atmospheric Sciences and Climate, and the Division on Earth and Life Studies. The committee identified the following as key roles for science and technology in accomplishing the goal of controlling air pollutants to minimize concentration levels to a minimal or acceptable risk to human health and welfare without "unduly disrupting" the national economy or its underlying technological infrastructure.

1. Quantifying risks to human health and public welfare (such as ecosystems) associated with varying concentrations, mixtures, and rates of deposition of air pollutants to establish air quality standards and goals.
2. Quantifying the source–receptor relationships that relate pollutant emission rates to ambient pollutant concentrations and deposition rates in order to develop air pollution mitigation strategies to maximize benefits and minimize costs.
3. Quantifying the expected demographic and economic trends with and without air pollution control strategies to better account for growth in activity that might offset pollution control measures and to better design control strategies that are compatible with the economic incentives of those who must implement them.
4. Designing and implementing air quality monitoring technologies and methods for documenting pollutant exposures to identify risks and set priorities.
5. Designing, testing, and implementing technologies and systems for efficiently preventing or reducing air pollutant emissions.
6. Designing and implementing methods and technologies for tracking changes in pollutant emissions, pollutant concentrations, and human health and welfare outcomes to document and ultimately improve the effectiveness of air pollution mitigation activities.

2 An Expanding Federal Presence in Air Quality Controls

The Clean Air Act amendments of 1970 (also referred to as "The Clean Air Act") dramatically reframed national, state, and industry relationships. Within months of enactment, the newly formed EPA designated permissible ambient air concentrations for six pollutants based on health impact studies. The agency followed by directing states to prepare implementation plans that would meet quality standards for carbon monoxide, ozone, lead, nitrogen oxides, particulate matter, and for sulfur dioxide. Operators of major source emitters would be required to meet control standards without regard for costs or currently available technologies.

Amendments in 1990 added a national permit program for toxic air pollutants to be administered and financed by state regulators. Ongoing state and industry initiatives have triggered Congressional debate over appropriate implementation strategies. The following sections focus on the adaptations by government administrators, and by industries with their respective trade associations, in relation to federal Clean Air Act legislation.

2.1 Precedent for the Clean Air Act of 1970

2.1.1 Pre-1955 History

Smoke and fog events have been linked to respiratory illnesses throughout the industrial era. Los Angeles cataloged its first smog incident in 1943 and created California's first Air Pollution Control District in 1947. Los Angeles created the state's first district in that year.[1] In 1948, residents of Donora, Pennsylvania, were impacted severely by a 4-day meteorological inversion that brought a persistent fog, laden with particulate and industrial contaminants (Photo 2.1). This event was linked to 20 deaths, 400 hospitalizations, and an estimated 6,000 reports of respiratory symptoms among the town's 14,000 residents.[2] The London "Great Killer Fog" in December 1952 was a 4-day air inversion that interacted with smoke and other

[1] South Coast Air Quality Management District, "The Southland's War on Smog: Fifty Years of Progress Toward Clean Air," May 1997. http://www.aqmd.gov/news1/Archives/History/marchcov.html (accessed October 2011).

[2] William H. Helfand, Jan Lazarus, and Paul Theerman, "Donora, Pennsylvania: An Environmental Disaster of the 20th Century," 91 *American Journal of Public Health* 553, 553 (2001); John Bachmann, "Will the Circle Be Unbroken: A History of the U.S. National Ambient Air Quality Standards," 57 *Journal of the Air & Waste Management Association* 652, 659 (2007).

An Interactive History of the Clean Air Act. DOI: 10.1016/B978-0-12-416035-4.00002-3

Photo 2.1 Donora, Pennsylvania, 1948. http://www.alleghenyfront.org/story.html?storyid = 200808191557080.465102 (accessed August 2011).
Source: Courtesy of *Pittsburgh Gazette*.

pollutants. This incident was considered the cause of 3,000 deaths during this period, and an estimated 13,000 more, over the following 6 months through attributable causes.[3]

2.1.2 Establishing the Federal Presence in Air Quality Management

The United States Air Pollution Control Act of 1955[4] marked the first direct national commitment to air quality issues. It charged the Public Health Service within the Department of Health, Education, and Welfare (DHEW) with coordinating information gathering and scientific findings relating to the causes and effects of air pollution.[5] The 1955 law also provided support for state and local technical assistance, training, and program development. This framework placed the federal government in an indirect supporting role via research grants to universities and technical assistance for local and state programs. Figure 2.1 illustrates this initial structure.

The Clean Air Act of 1963 established the first direct authority for national intervention in air pollution "which endangers the health or welfare of any person."[6] States could petition the DHEW to hold a hearing, convene a conference, or initiate a federal lawsuit if a polluting source, or sources, did not respond satisfactorily. The

[3] Davis, Devra. *When Smoke Ran Like Water*. Basic Books, 2002, pp. 78–9.

[4] The Air Pollution Act of 1955, Public Law No. 84–159 (1955), and 1960 amendments, supported research, and technical assistance to states and to local air pollution control districts. *See* Bachmann, note 2, at 660–62.

[5] Air Pollution Control Act of 1955. *See also* Martin A. McCrory and Eric L. Richards. "Clearing the Air: The Clean Air Act, GATT and The WTO's Reformulated Gasoline Decision." 17 *UCLA Journal of Environmental Law and Policy* 1, 4 (1999).

[6] Public Law No. 88–206 § 5(a) (1963).

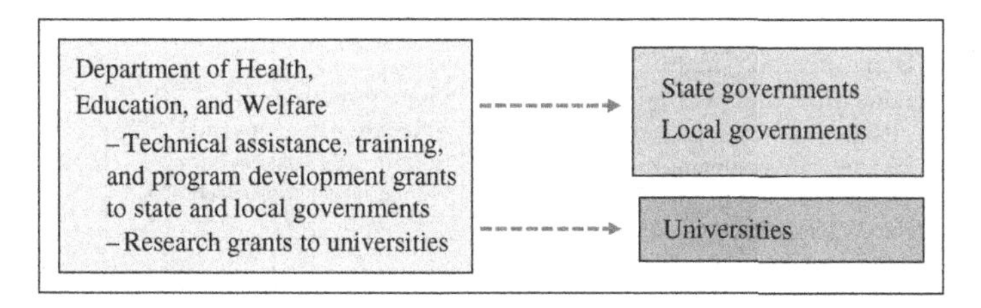

Figure 2.1 Air pollution control relationships: United States Air Pollution Act of 1955.

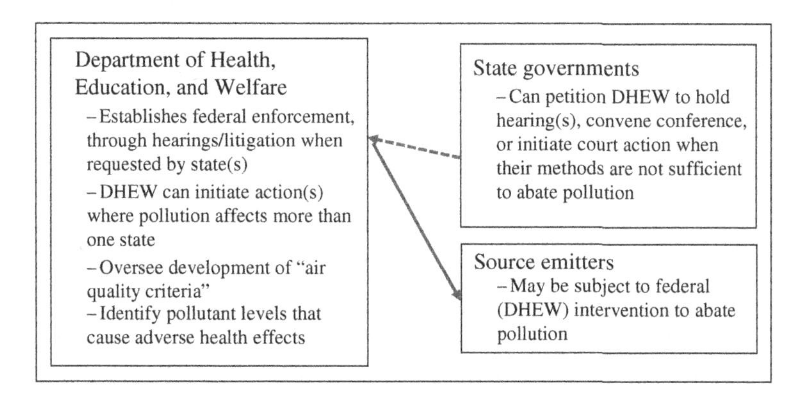

Figure 2.2 Elements added by Clean Air Act of 1963.

1963 law also empowered the DHEW to initiate actions where pollution issues from one state impact the quality of life for individuals in a neighboring state.[7] Another key provision directed the National Air Pollution Control Administration to oversee development of "air quality criteria" and to "identify pollutant levels that cause adverse health effects."[8] The 1963 law also served as the base for the major amendments is 1970 and 1990 (Figure 2.2).

The 1965 Motor Vehicle Air Pollution Control Act was the first to mandate federal standards for manufacturers. It authorized the DHEW to set standards for automobile emissions beginning with the 1968 model year.[9] It also prescribed that these changes be economically and technologically feasible.[10] The agency responded

[7] Public Law No. 86–493 (1960). *See* Arnold W. Reitze, Jr., "A Century of Air Pollution Control Law: What's Worked; What's Failed; What Might Work," 21 *Environmental Law* 1549, 1588–90 (1991).

[8] Public Law No. 88–206 (1963). *See also* Edmund S. Muskie, "NEPA to CERCLA: The Clean Air Act: A Commitment to Public Health." *The Environmental Forum.* January/February, 1990. The Clean Air Trust (2002). http://www.cleanairtrust.org/nepa2cercla.html (accessed October 2011); James R. Fleming and Bethany R. Knorr, "History of the Clean Air Act," *American Meteorological Society* 1999. http://www.ametsoc.org/sloan/cleanair/index.html (accessed October 2011); Jack Lewis, "The Birth of EPA," *EPA Journal* (November 1985).

[9] Motor Vehicle Pollution Control Act (1965) 79 Stat. 992 (1965). *See* Robert Martin and Lloyd Symington, "A Guide to the Air Quality Act of 1967," 33 *Law and Contemporary Problems* 239, 241–44 (1968). *See also* McCrory and Richards, *supra*, note 5.

[10] Motor Vehicle Pollution Control Act (1965) § 202(a).

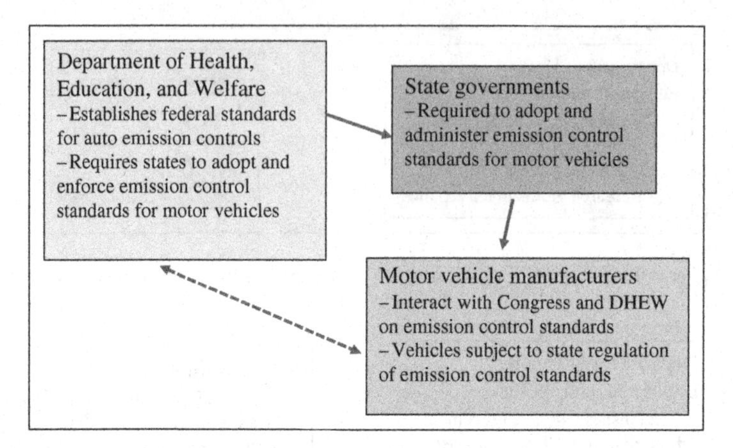

Figure 2.3 Motor Vehicle Air Pollution Control Act of 1965.

by adopting the levels set by the State of California for the 1966 model year.[11] California had used the 1963 model year as its base for reducing hydrocarbons by 72%, carbon monoxide by 56%, and for eliminating emissions of crankcase hydrocarbons altogether.[12] States were now required to regulate vehicles based on either the contemporary federal standards or those enacted by the State of California for control of mobile emissions. Figure 2.3 illustrates this added element.

2.1.3 Toward a Federal–State Cooperative Framework: The Air Quality Act of 1967

The Air Quality Act (AQA) of 1967 began the transition toward the current federal–state structure for planning and regulation. It required states to adopt ambient air quality standards and prepare implementation plans for reducing pollution to acceptable levels. The AQA also affirmed the mandate for states to regulate automobile emissions according to California or federal standards. However, the 1967 law reiterated that pollution control programs and enforcement were clearly within each state's jurisdiction.[13]

One key provision of the AQA authorized the DHEW Secretary to determine air quality criteria " … as in his judgment may be requisite for the protection of the public health and welfare."[14] The agency responded by funding research to measure the impacts of incremental exposure to pollutants that would affect lung capacity in at-risk individuals. The findings from these controlled studies identified threshold

[11] *See* Harold W. Kennedy and Martin E. Weekes, "Control of Automobile Emissions—California's Experience and the Federal Legislation," 33 *Law and Contemporary Problems* 297 (1968).

[12] *See* "Mobile Sources: Emissions from Automobiles, Trains, and Airplanes," *American Meteorological Association* (2002). http://www.ametsoc.org/sloan/cleanair/cleanairmobile.html#caamost (accessed October 2011).

[13] Arnold W. Reitze, Jr., "The Role of the 'Region' in Air Pollution Control," 20 *Case Western Reserve Law Review* 809 (1969).

[14] Air Quality Act of 1967, Public Law No. 90–148, § 107(b)(1)(1967).

concentrations in ambient air that were determined to diminish those subjects' health. The epidemiological data from this research became the scientific bases for "criteria documents" underlying EPA's initial determination of NAAQS.

Unfortunately, states were not bound by the DHEW criteria, which brought the potential for 50 separate sets of air quality standards.[15] Further, because the AQA did not require state plans to be submitted for federal review, there was limited success in this voluntary approach. By 1970, only 21 implementation plans had been submitted by 15 states. California had formulated plans for Metropolitan Los Angeles and the San Francisco region. There was a Connecticut plan for Hartford–New Haven–Springfield, a Massachusetts plan for the Metropolitan Boston Region, a Colorado plan for the Metropolitan Denver Region, a Maryland plan for Metropolitan Baltimore, and the Missouri plan for Metropolitan St. Louis. There were also interstate plans for the National Capital Region (District of Columbia, Maryland, and Virginia), Metropolitan Chicago (Illinois and Indiana), the tristate New York region (New York, New Jersey, and Connecticut), and Metropolitan Philadelphia (Pennsylvania, Delaware, and New Jersey), and the Kansas City region (Kansas and Missouri).[16]

Another AQA requirement directed the DHEW to designate of Air Quality Control Regions, also referred to as "air basins" or "air sheds," in consultation with state and local jurisdictions.[17] DHEW guidelines recommended that these regions be established based on common characteristics such as degree of urbanization, industrial development, jurisdictional boundaries, pollution issues, and emission types considered "relevant to effective implementation of air quality standards."[18]

Figure 2.4 shows the governmental relationships established by the AQA. States, substate, and contiguous interstate Air Quality Control Regions were at the center of this strategy. Federal standards could be applied only to situations where air pollution issues transcended state boundaries without adequate plans for compliance. Neither the states nor DHEW was subject to legislatively imposed time restraints.

Despite its limited program success, the AQA provided a foundation for the Congressional policy that would emerge in 1970. The DHEW's administration established key precedent for basing policy and program development on underlying scientific studies. The AQA also ensured that states would retain their central implementation role in planning and regulation. Federal lawmakers would continue to rely on scientific advances and technological innovations as the cornerstones of cooperative air quality management. In practice, industrial representatives would interact

[15] Steve Norvick and Bill Westerfield, "Whose SIP Is it Anyway? State–Federal Conflict in Clean Air Act Enforcement."18 *William and Mary Journal of Environmental Law* 245, 260–61 (1994).

[16] EPA, "EPA Approves First State Clean Air Plans." Press release, August 13, 1971. http://www.epa.gov/history/topics/caa70/12.html (accessed October 2011).

[17] See Arnold W. Reitze, Jr., "Air Quality Protection Using State Implementation Plans—Thirty-Seven Years of Increasing Complexity," 15 *Villanova Environmental Law Journal* 209, 212 (2004); James A. Holtkamp, *The Clean Air Act*, 1–2. Holland & Hart LLP, August 2003. *See also* Lisa Heinzerling, "Ten Years After the Clean Air Act Amendments: Have We Cleared the Air?" 20 *St. Louis University Public Law Review* 121, 134–35 (2001).

[18] Sussman, V.H., "New Priorities in Air Pollution Control," *Conference on Air Pollution Meteorology*, 1–2. American Meteorological Society, 1971. http://www.ametsoc.org/Sloan/cleanair/pdfdocs/sussman.pdf (accessed October 2011).

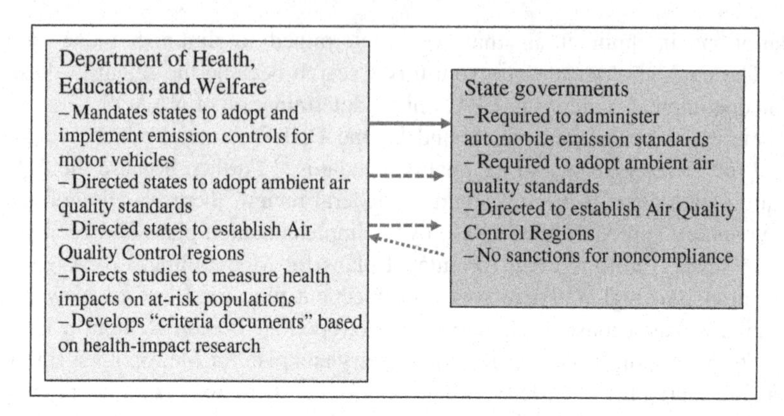

Figure 2.4 AQA of 1967 reframes federal–state relationships.

continually with national standard-setters and state regulators. EPA, state agencies, and corporate environmental programs would adjust concerns over feasibility, costs, and timing for pollution control technologies.

2.2 The Framework of the Clean Air Act of 1970

The 1970 Clean Air Act amendments marked three major shifts in pollution prevention strategies. First, it changed the national emphasis from facilitating state programs to establishing and enforcing NAAQS. This element directed the newly formed EPA to identify pollutants and their sources, and to establish criteria for levels that ensure public health and welfare. While continuing to delegate primary implementation to states, the 1970 amendments required each to submit a timely implementation plan to the EPA directed toward meeting federal criteria. The Clean Air Act's third unprecedented element acknowledged that existing control technologies could be inadequate to meet federally established standards. If so, industries would need to develop pollution control technologies for both stationary and mobile sources.

In proposing these amendments, Senator Edmund Muskie (Democrat, Maine) highlighted his concerns with the AQA and its predecessors. Foremost, he cited a perceived confusion about varying state standards for acceptable air quality. Muskie explained further that voluntary compliance and unlimited time frames had proven to be ineffective strategies. An additional realization was that existing control technologies were not significantly reducing emissions from stationary and mobile sources. Muskie implored his Senate colleagues to consider that:

> *A nation which has been able to conquer the far reaches of space, which has unlocked the mysteries of the atom, and which has an enormous reserve of economic power, technological genius, and managerial skills, seems incapable of halting the steady deterioration of our air, water, and land.*[19]

[19] EPA, *Legal Compilation 1973*, Vol. 2, Statutes and Legislative History, "Amendment of Clean Air Act." Remarks by Senator Edmund M. Muskie, at p. 1493.

He followed by emphasizing the preeminent goal of protecting public health even when there are no presently apparent means for accomplishing it:

> *... The first responsibility of Congress is not the making of technological or economic judgments—or even to be limited by what is technologically or economically feasible. Our responsibility is to establish what the public interest requires to protect the health of persons. This may mean that people and industries will be asked to do what is impossible at the present time ....*[20]

Senator Muskie described five elements essential to implementing the law: (1) identifying pollutants and compliance criteria to protect public health; (2) regional implementation within a specified time frame (initially 5 years); (3) ensuring that new facility construction would not degrade air quality in areas that already comply with national standards; (4) establishing national authority to regulate toxic emissions; and (5) establishing additional federal authority to regulate pollutants not covered by NAAQS or new source standards.[21] His then Chief of Staff highlighted the bill's citizen enforcement provisions as key to its implementation success.[22]

Cosponsor Howard Baker (Republican, Tennessee) recollected the following elements as key to the Senate's unanimous adoption. First, there was consensus on "a direct and overarching federal interest in protecting the health of all Americans from air pollution ...." This incorporated the concept of technology forcing and establishing mandates and deadlines for government actions. The 1970 amendments also empowered individual American citizens "with the authority to use the federal courts to achieve the objectives we set forth should the bureaucracy or the politicians fail to do so."[23] Senator Baker expressed a belief that the public could rightfully rely on their elected officials to balance the capacity to abate pollution against the health risks from failing to achieve those reductions.[24]

> *We respected the technical, scientific and regulatory skills that were available to the Federal Government and to the states. But at the end of the day judgments with respect to the availability of technology—the costs of pollution control—and the ability to meet standards by specific deadlines were political, not bureaucratic judgments.*[25]

[20] Ibid., at pp. 1496–97.

[21] Ibid., at p. 1497.

[22] Leon Billings, "The Muskie Legacy: Policy and Politics," April 14, 2005. http://www.muskiefoundation.org/billings.lecture.041405.html (accessed October 2011). Billings stated that: ... While many look at the introduction of the concept of mandatory requirements, statutory standards, statutory deadlines, establishment of scientifically based public health standards or even the removal of economic and technical feasibility tests and other weasel words to be the singular most significant aspects of the Clean Air Act, in my view the thread that holds all of these provisions together is the right of citizens ... to force government agencies to do the job the law requires. Ibid.

[23] Howard H. Baker, "Cleaning America's Air—Progress and Challenges." Remarks by Howard H. Baker, Jr., March 9, 2005. http://www.muskiefoundation.org/baker.030905.html (accessed October 2011).

[24] Ibid.

[25] Ibid.

The political compromise leading to the 1970 Clean Air Act amendments brought agreement on "science-based air quality standards that could force development of emission control technologies" while allowing the EPA Administrator to grant limited extensions.[26]

From its inception, the Clean Air Act of 1970 integrated federal public health standards, state-based compliance programs, and demands for industrial innovation.[27] Its primary pollution prevention goal is to "encourage or otherwise promote reasonable Federal, State, and local governmental actions" consistent with its provisions.[28] Findings note that industrial development, population concentrations within metropolitan areas, and the increasing use of motor vehicles[29] bring "mounting dangers to the public health and welfare."[30] These negative impacts extend to "agricultural crops and livestock," as well as property damage, and endangerment of air and ground transportation.[31] The Act acknowledges that "the amount and complexity of air pollution"[32] are compounded by population growth and concentrated development patterns that are not limited by state and local boundaries.[33] However, it affirms that these governments have primary responsibility for the prevention of air pollution from both stationary and mobile sources.[34]

2.3 Allocating Administrative Roles to Meet Clean Air Act Requirements

The federal executive branch responded to Clean Air Act requirements by consolidating leadership within the EPA.[35] While it continued the structure of state-based implementation plans, these would now be subject to federal certification for conformance with the NAAQS. States would be required to regulate automobile emissions according to either EPA standards or more to stricter criteria established by the State of California. However, implementation plans could vary requirements for stationary sources provided they meet overall statewide criteria for ambient air quality. In other words, a petrochemical plant in New Jersey might be subject to different requirements than a comparable facility in Oklahoma. Regulations could also vary from location to location within a state based on area-wide pollutant levels.

As initially envisioned by the Clean Air Act and the EPA, SIPs would reach compliance with the NAAQS by 1975. Further, automotive manufacturers would meet emission control standards by this date. New or expanded state agencies would be

[26] Ibid.
[27] Clean Air Act, § 101(a)(4). *See also* Muskie, "NEPA to CERCLA," note 8.
[28] Clean Air Act § 101(c).
[29] Clean Air Act § 101(a)(1)–(2).
[30] Clean Air Act § 101(a)(2).
[31] Clean Air Act § 101(a)(2).
[32] Clean Air Act § 101(a)(3).
[33] Clean Air Act § 101(a)(1).
[34] Clean Air Act § 101(a)(3).
[35] Paul Rogers, "The Clean Air Act of 1970," *EPA Journal* (January/February 1990); James E. McCarthy, *Clean Air Act: A Summary of the Act and its Major Requirements*. Congressional Research Service, May 2005.

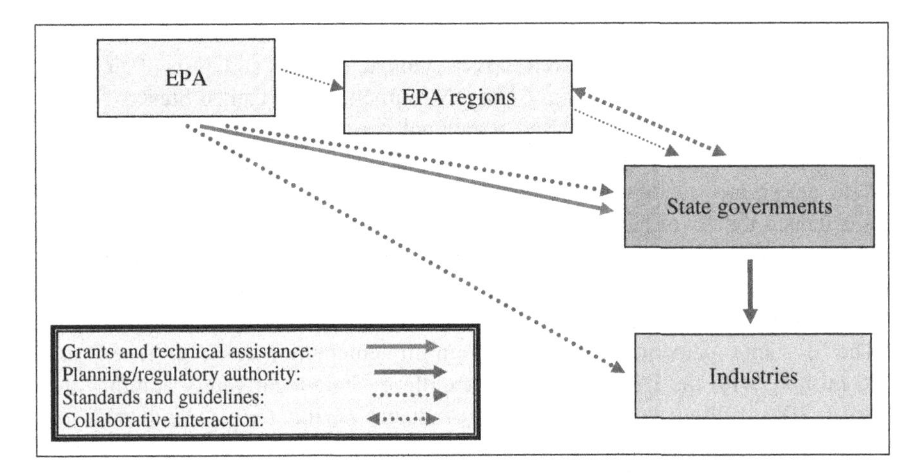

Figure 2.5 Initial structure of the Clean Air Act of 1970.

responsible for monitoring air quality, regulating pollution sources, and communicating directly with representatives for those regulated industries. The 1970 amendments conveyed optimism that American industries could improve air quality by developing new control technologies.

Figure 2.5 illustrates the interactive structure in the initial legislative design. The EPA provides both guidance and regulatory review authority toward states. Primary interactions with state-level programs are via the agency's 10 regional offices. State programs carry out planning and implementation using their inherent regulatory authority. In order to meet the NAAQS, regulations and other controls may be more stringent in selected nonattainment areas. Industries interact with both state officials and through advocacy with federal legislative and executive representatives and through judicial processes.

In 1977, Congress adapted the requirements and time frames for meeting state and industry requirements. Compliance dates for SIPs would be deferred. Other provisions clarified that air quality in areas presently meeting the NAAQS could not be compromised in order to meet statewide standards. The 1977 amendments sought to further define the standards for industrial technologies for pollution control. Initially, EPA determined the Best Available Control Technology (BACT) to reduce emissions from stationary sources. Over time, criteria for industry compliance would allow for Reasonably Available Control Technology (RACT), Lowest Achievable Emission Rates (LAER), and National Emission Standards for Hazardous Air Pollutants (NESHAP).[36]

2.4 The 1990 Clean Air Act Amendments Redefine Federal, State, and Industry Roles

The 1990 Clean Air Act amendments placed additional mandates on EPA, states, and businesses. These included industry-specific requirements for control technologies,

[36] *See* Holtkamp, note 17.

new standards for mobile source emissions, and programs to eliminate toxic air pollutants.[37] Another provision required EPA to evaluate benefits and costs of air regulations on the "public health, economy, and environment of the United States."[38]

The 1990 amendments authorized a national permit program for 189 toxic pollutants specified within its provisions. Codified as "Title V," the revised law directed EPA to determine standards and regulations for major and minor source emitters. This extended the net of federal regulation to include businesses such as dry cleaners, print shops, paint dealers, and restaurants. Title V federal permits would be administered by individual states. These state programs would be required to be financially self-supporting through fees, enforcement actions, and other revenue sources.

The EPA anticipated total annual program implementation costs for Title V to be $526 million, with the EPA's costs at $14 million. State/local yearly administrative costs of $160 million would be offset by industry permit fees. Overall additional industry costs were estimated at $352 million/year.[39] Other provisions authorized sanctions for failure to either submit or adequately implement an SIP.

Private sector impacts of the Title V program extended beyond major industries to include emissions from businesses unfamiliar with federal and state environmental regulations. New trade associations emerged to assist dry cleaners, print shops, paint retailers, and other businesses that would require new technologies, equipment, and operating procedures. Interstate coalitions supported policy analysis and strategies to address pollution issues that transcended their borders. The 1990 amendments continued to reflect EPA's shift toward collaborative policy development. The agency sought direct input from trade associations, states, and other sources via advisory committees and policy negotiations on pending regulations.

2.5 Clean Air Act Strategies: Air Quality Management and Direct Standards for Stationary and Mobile Sources

Clean Air Act implementation merges a state-controlled "air quality management" approach[40] with federal regulations to control pollution at its sources. Both rely on scientific findings and engineering technologies for policy development. EPA's health-based NAAQS set the overall criteria for statewide ambient air concentration limits. Detailed implementation planning within substate regions determines mix of regulations and development policies that will attain or maintain these levels.

The second strategy applies direct federal regulatory standards for stationary and mobile sources. Within this context, EPA can define available or prototype control technologies applicable to industrial sites, motor vehicles, and businesses

[37] Congressional Research Service, *Clean Air Act: A Summary of the Act and its Major Requirements*, note 35, at p. 10.

[38] See § 812 of the Clean Air Act Amendments of 1990.

[39] Title V Task Force, *Final Report to the Clean Air Act Advisory Committee: Title V Implementation Experience*, at p. 20 (April 2006).

[40] Bachmann, note 2, at pp. 652–54.

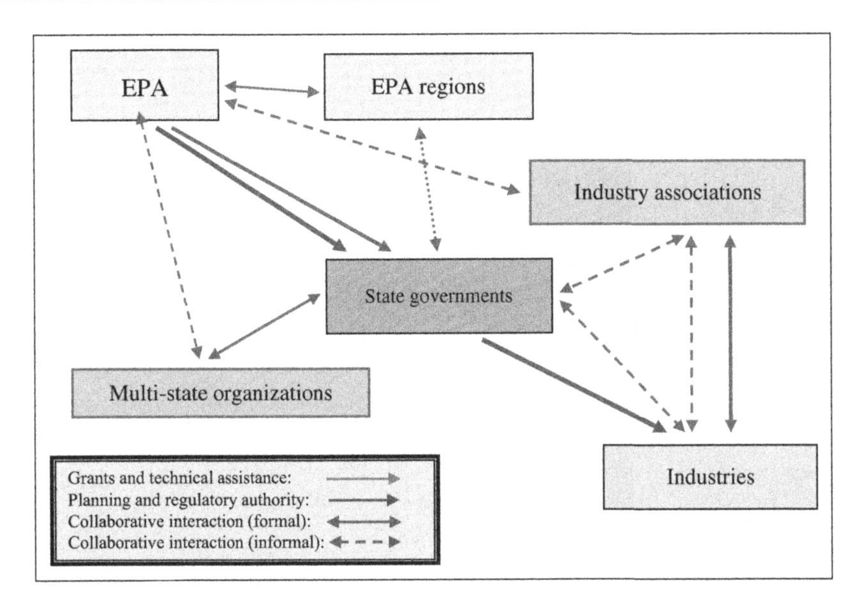

Figure 2.6 Current structure for Clean Air Act Interactions.

that emit identified toxic pollutants. These standards are determined typically through negotiations between the agency and representatives for affected interests. Once adopted, states must ensure that applicable technologies and practices are implemented.

2.6 Current Dynamics in Clean Air Act Implementation

Current policy dynamics in Clean Air Act implementation intertwine EPA, state, industry, and other stakeholders. The EPA maintains formal relationships with states on SIP review and for toxic emission regulations. Industry trade associations represent their sectors as advocates and collaborators in policy processes. The EPA's regional offices act as intermediaries between the Washington office and the state program authorities. State and industry consortiums continue to trigger debate over appropriate implementation strategies. Figure 2.6 depicts the complexity of these concurrent interactions. The following chapters focus on the adaptations by federal and state administrators, and by industries and their respective trade associations in relation to federal Clean Air Act legislation.

3 Federal Leadership in Clean Air Act Implementation: The Role of the Environmental Protection Agency

3.1 Phase I: 1970–1977

3.1.1 Establishing the Clean Air Act's Administrative Structure

President Richard M. Nixon's Advisory Council on Executive Organization proposed establishment of an independent agency charged with environmental protection.[1] The Council recommended that standard-setting functions supported by strong research capacities should be conducted outside the scope of existing agency interests. It reasoned further that an independent executive agency would reduce state and local government "forum shopping" for assistance among multiple agencies. The single-agency approach would also "simplify the relationship of the private sector whose cooperation and ingenuity are essential if any real progress is to be made."[2]

In a July 9, 1970, Special Address to Congress, the President highlighted the need for an executive agency to serve as a center point for environmental research, monitoring, standard-setting, enforcement, and state/local assistance.[3] It would identify pollutants, how they interact with respect to human exposure, and "where in the ecological chain interdiction would be most appropriate …."

> *Despite its complexity, for pollution control purposes the environment must be perceived as a single, interrelated system. Present assignments of departmental responsibilities do not reflect this interrelatedness.*[4]

An Executive Order established the EPA on December 2, 1970[5] (Photo 3.1). The President promised that states would have opportunities to make a "good faith effort"

[1] Executive Office of the President, President's Advisory Council on Executive Organization, "Memorandum for the President, Subject: Federal Organization for Environmental Protection," April 29, 1970. *See generally* Bachmann, "Will the Circle Be Unbroken," at pp. 671–76; Lyle Witham, *A Summary of the Development of the Clean Air Act* 3–11. North Dakota Office of the Attorney General, August 2005.

[2] EPA, "History: Ash Council Memo. Subject: Federal Organization for Environmental Protection," *Executive Office of the President, President's Advisory Council on Executive Organization, Memorandum for the President.* April 29, 1970.

[3] President Richard M. Nixon, "Special Message from the President to the Congress About Reorganization Plans to Establish the Environmental Protection Agency and the National Oceanic and Atmospheric Administration, Reorganization Plan No. 3 of 1970." July 9, 1970.

[4] Ibid.

[5] Ibid.

An Interactive History of the Clean Air Act. DOI: 10.1016/B978-0-12-416035-4.00003-5

Photo 3.1 Appointment ceremony for William Ruckelshaus as first EPA Administrator. *Source*: From National Archives and Records Administration RG412. http://earth911.com/news/2010/06/28/how-america-has-gone-green (accessed August 2011).

toward implementing air quality standards.[6] He warned that federal enforcement would be "swift and sure" where state regulatory controls were insufficient to protect environmental degradation.

The EPA's initial organizational structure positioned the Air Pollution Control Office at a parallel level with water, pesticides, radiation, and solid waste. It was formed by consolidating programs from the DHEW and other executive agencies.[7] The Air Office's broad charges included identifying acceptable levels of pollutants "… required to minimize or eliminate deleterious effects," and overseeing "… the achievement of a wholesome air environment through development of air pollution control technology."[8] These objectives would be implemented through regulatory programs for stationary and mobile source emissions. The Office would also administer research and development programs, federal–state–local air quality monitoring, and programs for technical assistance, training, and financial support.[9]

[6] Jack Lewis, "The Birth of EPA," *EPA Journal* (November 1985).

[7] The DHEW National Air Pollution Control Board brought a staff of 1,020 and $81.4 million annual budget. The Environmental Control Administration within the Public Health Service had directed the scientific studies that provided the bases for establishing national ambient air quality standards. *See* EPA, "History: Ash Council Memo. Subject: Federal Organization for Environmental Protection," *Executive Office of the President, President's Advisory Council on Executive Organization, Memorandum for the President.* April 29, 1970.

[8] EPA, "Environmental Protection Agency, ORDER 1110.2, Initial Organization of the Environmental Protection Agency," December 4, 1970.

[9] Ibid, Section 9.

Within 2 weeks, Administrator William Ruckelshaus established 10 multi-state regional offices[10] (Figure 3.1) This framework was consistent with President Nixon's "New Federalism" strategy to decentralize national authority. An underlying rationale was that Regional Administrators could act as "cutting edge" liaisons between the EPA's Washington staff and industries, private organizations, states, and municipalities.[11] They could collect pollution information and refer offenders to the Justice Department for prosecution. This direct line of authority between the Washington and regional offices was intended to provide a balance between a top–down command and control approach and ensuring that states would retain primary implementation responsibilities.[12] Reflecting on these early experiences, Administrator Ruckelshaus explained that this structure enabled federal authority to be unleashed as a "gorilla in the closet" where "state authorities were either too weak or too inept to control local polluters ..."[13] (Figure 3.2).

The EPA's resolve was tested on January 9, 1971, when it ordered Union Carbide to bring an Ohio plant into compliance with recommended standards for sulfur oxide emissions.[14] The company responded by threatening to lay off 675 workers. After

[10] The Region 1 office in Boston acts as liaison between the Washington office and the New England states of Connecticut, Maine, Massachusetts, New Hampshire, Rhode Island, Vermont, and 10 tribal nations. It is characterized by dense population areas, an aging industrial base, and undisturbed natural areas. EPA Region 2 covers New Jersey, New York, Puerto Rico, U.S. Virgin Islands, and seven tribal nations. Nearly 85% of its 31 million residents live in New York State. New Jersey is the region's most heavily populated state. This region includes several unique ecological areas, including the New Jersey Pine Barrens, Adirondack State Park in New York, the Hudson River, the Caribbean National Forest, and the Virgin Islands National Park. The Mid-Atlantic Region (Region 3) is based in Philadelphia and covers the states of Delaware, District of Columbia, Maryland, Pennsylvania, Virginia, and West Virginia. EPA Region 4 covers eight states (Alabama, Florida, Georgia, Kentucky, Mississippi, North Carolina, South Carolina, and Tennessee), and seven tribal nations. Its east–west boundaries range from the Atlantic Ocean to the Mississippi River. Illinois, Indiana, Michigan, Ohio, Minnesota, and Wisconsin, and 35 tribal nations are within EPA's Great Lakes Region (Region 5). It includes long-established steel, automobile, and other manufacturing industries. Region 6 covers Arkansas, Louisiana, New Mexico, Oklahoma, and Texas. This five-state region is economically and ecologically diverse. More than one-half of Arkansas's landscape is predominately forested and aquatic. Texas is the region's most populous state. Region 7 interacts with the midwestern states of Iowa, Kansas, Missouri, and Nebraska. Colorado, Montana, North Dakota, South Dakota, Utah, Wyoming, and 27 sovereign nations are within EPA Region 8. The Rocky Mountains, Great Plains, and Colorado Plateau are among its natural areas. About two-thirds of its 10 million residents live in developed regions along Colorado's Front Range and Utah's Wasatch Front. The Pacific Southwest states of Arizona, California, Nevada, as well as Hawaii and 147 recognized tribes comprise EPA Region 9. The Pacific Northwest Region (Region 10) includes Alaska, Idaho, Oregon, Washington, and approximately 270 tribal governments.

[11] William D. Ruckelshaus, "First Administrator on Establishment of EPA." Press Release, December 16, 1970. *See* Clean Air Act § 301(a)(2). Dennis C. Williams, "The Guardian: EPA's Formative Years, 1970–1973," September 1993. http://www.epa.gov/aboutepa/history/publications/print/formative.html (accessed October 2011).

[12] *See* "William D. Ruckelshaus: Oral History Interview," conducted by Michael Dorn, January 1993. http://epa.gov/aboutepa/history/publications/print/formative.html (accessed October 2011).

[13] Ibid.

[14] Ibid.

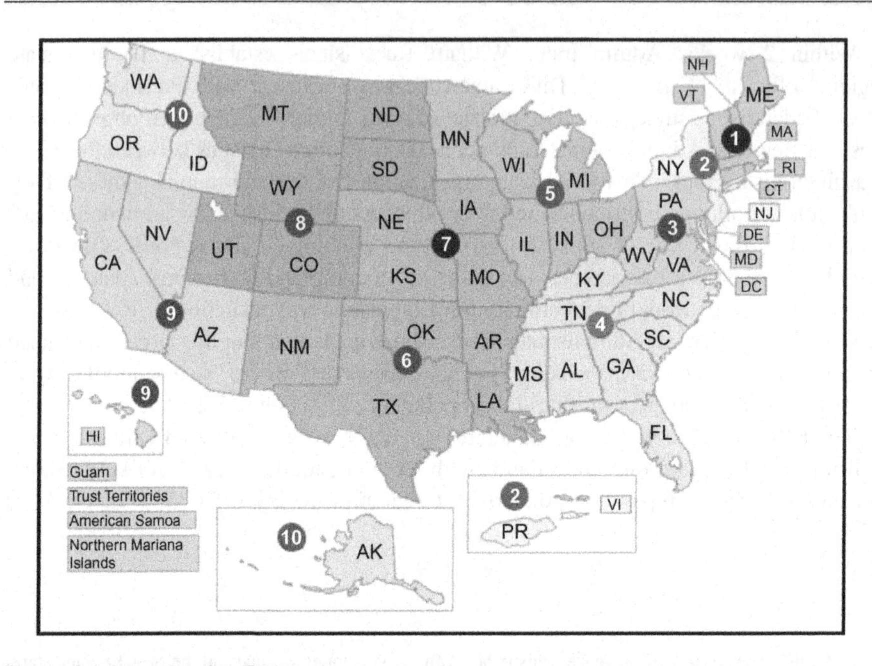

Figure 3.1 Map showing EPA regions.
Source: EPA, "About Regions." http://www.epa.gov/epahome/locate2.htm (accessed October 2011).

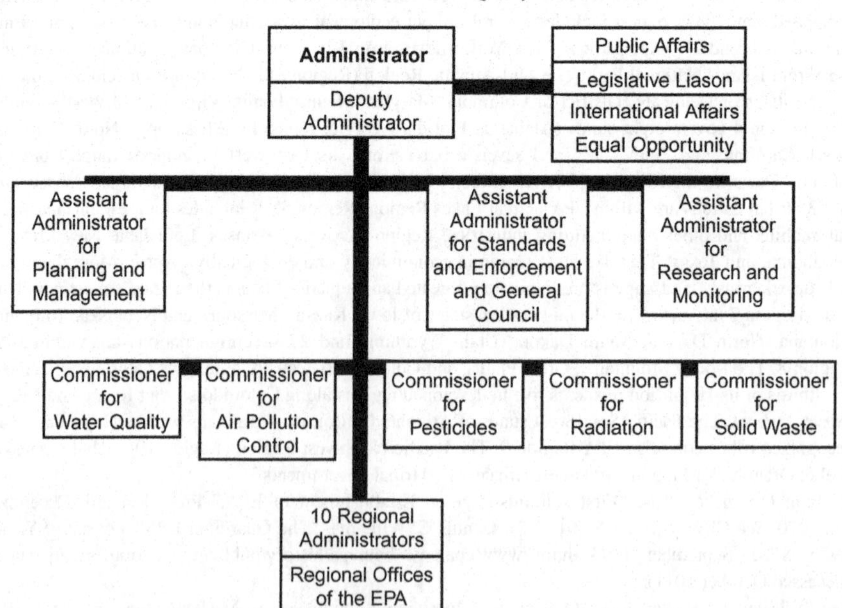

Figure 3.2 The EPA's initial organizational structure.
Source: EPA, "EPA organization, December 15, 1970." http://www.epa.gov/history/ images/figure2.gif (accessed October 2011).

extended negotiations, Union Carbide agreed to reduce emissions according to an expedited schedule.[15]

Interactions among the regional and central offices raised frustration among states and industries on ultimate authority for policy and regulatory decisions. Douglas M. Costle, Administrator from 1977 to 1981, noted an inherent tension facing regional offices as intermediaries between Washington and state regulators:

> *You want the regions to be close to the States, to give the day-to-day attention to state programs that headquarters can't. But if you just turn all ten regions loose, you'd have chaos. So you always need a balance between national guidelines and enough flexibility to meet the problems in the field.*[16]

3.1.2 Establishing Air Quality Standards and a Framework for State Implementation

One of the EPA's primary initial tasks was to identify pollutants and permissible levels that would meet the Act's goals to protect public health and public welfare.[17] Agency scientists, aided by an advisory panel, developed NAAQS for six "criteria pollutants": carbon monoxide (CO), ozone (O_3), lead, nitrogen oxides (NO_x), particulate matter (10 or less microns in diameter), and sulfur dioxide (SO_2). These standards incorporated findings from epidemiological studies authorized by DHEW under the 1967 Air Quality Act.

Administrator Ruckelshaus affirmed that the initial standards were "... based on investigations conducted at the outer limits of ... [the agency's] capability to measure connections between levels of pollution and effects on man."[18] He indicated that they were designed to include protections for persons with preexisting cardiorespiratory problems. "*If we have erred at all in setting these standards, we have erred on the side of public health*" [emphasis added].[19] The initial designations in April 1971 set the stage for controversies over the relationship between policy determinations and their scientific underpinnings.

This initial push to meet federal mandates was problematic for states and affected industries. Ruckelshaus described his challenges in assisting states in the following way:

> *One of the first things I did at EPA was travel around the country and talk to state regulatory officials. I convened meetings with them in the various EPA regions.*

[15] Ibid.

[16] EPA, Douglas M. Costle, "Oral History Interview." http://nepis.epa.gov/Exe/ZyPURL.cgi?Dockey=1000494F.txt (accessed October 2011).

[17] *See* James A. Henderson and Richard N. Pearson. "Implementing Federal Environmental Policies: The Limits of Aspirational Commands," 78 *Columbia Law Review* 1429, 1430–31 (1978). *See also* Richard B. Stewart, "Pyramids of Sacrifice? Problems of Federalism in Mandating State Implementation of National Environmental Policy." 68 *Yale Law Journal* 1196, 1211–68 (1977).

[18] "EPA Sets National Air Quality Standards," EPA Press release, April 30, 1971.

[19] Ibid.

> *I heard the same story over and over and over again: "You're pushing us around too much; you're trying to dictate what ought to happen; we can handle this stuff ourselves; just give us more money, more federal grants; stay out of our hair." ... I had some quite angry meetings with these state regulatory officials.*[20]

Predictably, industrial groups and individual companies opposed the initial NAAQS rule.[21] One business leader introduced Ruckelshaus at a Washington event as "the greatest friend of American industry since Karl Marx."[22]

In this initial period, the EPA relied on models and extrapolations to anticipate the effects of air pollutants at lower levels and on limited knowledge "about the synergistic effect of ... what they might do in combination to public health or the environment." Ruckelshaus acknowledged the heuristic character of this approach, and expressed concern that the scientific standards relied on "... investigations conducted at the outer limits of our capability to measure connections between levels of pollution and effects on man."[23] These factors continue as challenges inherent in translating scientific findings into accurate and realistic policy directives.

After two years of implementation, the Administrator framed agency progress within a broader context of environmental awareness and legislation:

> *The real significance of the debate over individual environmental issues in the Agency's first year does not lie in the specifics not in the disposition of each case. The significance lies in the debate itself. The fact that it occurred—that questions never before asked about our personal and corporate actions were raised is a new, and I think, permanent element in American life.*[24]

The agency faced added challenges over implementation costs, uncertain time frames to control emissions, and educating citizens about air quality impacts from criteria pollutants.[25]

3.1.3 Defining Clean Air Act Parameters Through Litigation

During the 1970s, litigation was central to defining the scope and responsibilities in Clean Air Act implementation.[26] Federal courts established parameters for national and state discretionary authority as well as the extent that regulations could force

[20] William Ruckelshaus, "Oral History Interview," *supra* note 12.

[21] EPA received almost 400 comments in response to its proposed rule. *See* Richard H.K. Vietor, "The Evolution of Public Environmental Policy: The Case of 'Non-Significant' Deterioration," *Environmental Review* 3 (Winter 1979).

[22] "Environment: Ruckelshaus' First Year," *Time*, January 3, 1972. http://www.time.com/time/magazine/article/0,9171,879037,00.html (accessed July 2011).

[23] William Ruckelshaus, "EPA Press Release," April 30, 1971.

[24] EPA, *Progress Report, 1970–1972*, at p. ii.

[25] William Ruckelshaus, "Oral History Interview," *supra* note 12.

[26] *See* Samuel A. Bleicher, "Economic and Technical Feasibility in Clean Air Act Enforcement Against Stationary Sources," 89 *Harvard Law Review* 316 (1975).

industries to develop control technologies. Judicial and administrative decisions also affirmed that the Act requires states to protect air quality in areas that are within the established standards for criteria pollutants.[27]

The agency's second Administrator, Russell Train (1973–1977), predicted in 1974 that threshold disputes among powerful and legitimate public and private sector interests would be resolved beyond direct agency authority.

> *Our administrative, judicial and political processes now have the task of resolving these conflicts. They must do so by weighing all the interests which are affected in a sensitive and informed manner. Quick access to the legal dimensions of these problems is essential if conflicts are to be efficiently and fairly resolved.*[28]

The 1975 Supreme Court decision *Train v. Natural Resources Defense Council* upheld both the mandate for federally imposed air quality standards and the flexibility for states to vary their implementation plans. It acknowledged that the 1970 amendments were intended to remedy a voluntary-compliance approach by "taking a stick to the States ...."[29] However, the Court refused to affirm EPA authority to preempt state decisions if they are part of a timely plan to meet Clean Air Act standards.[30] This reasoning supported Georgia's discretion to offer variances within its state plan. However, the court did not rule on the question whether the state could allow facilities in nonattainment areas to disperse pollutants by constructing higher stacks as a means to meet overall compliance for the broader region.[31] This potential "tall-stacks" approach continued as a controversy in Clean Air Act implementation. Nonetheless, *Train v. Natural Resources Defense Council* remains significant for its support of federal standards enforced through state regulatory strategies.

In another controversial opinion, the Supreme Court supported the full authority of states to regulate stationary sources. The 1976 decision in *Union Electric v. EPA*[32] affirmed the Clean Air Act's "technology-forcing" provisions (Section 110) by ruling that Missouri's SIP could not be overturned "on the ground that it is economically or technologically infeasible."[33]

Legal processes also affirmed the interpretation that states must protect areas that already meet NAAQS standards. The Sierra Club brought suit in Federal District Court in 1972 claiming that the EPA needed to disapprove plans that allow pollution to increase in these areas as a means to attain overall air quality in a broader region.[34] The agency

[27] *Sierra Club v. Ruckelshaus*, 344F. Suppl. 253, at p. 256 (1972).

[28] Russell Train, "Foreword," in Environmental Protection Agency, *1974 Legal Compilation*, at pp. iii–iv.

[29] *Train v. Natural Resources Defense Council*, 421 U.S. 60, 64 (1975).

[30] Ibid., at p. 64.

[31] Ibid., at p. 79.

[32] *Union Electric v. EPA*, 427 U.S. 246 (1976). *See note*, "Forcing Technology: The Clean Air Act Experience," 88 *Yale Law Journal* 1713–14 (1979).

[33] *Union Electric v. EPA*, 427 U.S. at pp. 265–66. *See generally* Eric B. Easton and Francis J. O'Donnell, "The Clean Air Act Amendments of 1977: Refining the National Air Pollution Control Strategy," 27 *Journal of the Air Pollution Control Association* 943–47 (1977).

[34] *See Sierra Club v. Ruckelshaus*, 344F. Suppl. 253, at p. 256.

responded initially to this challenge by contending that the Clean Air Act did not prohibit "degradation of clean areas" and lacked authority to direct state actions in this area.[35]

The Supreme Court sustained the trial court's rejection of the agency's interpretation[36] in 1973.[37] In the interim, EPA adopted "nondegradation" regulations clarifying that SIPs must protect against significant deterioration of air quality in attainment areas.[38] This EPA rule required states to designate geographic areas as either attainment or nonattainment, and then to integrate specific plans for nonattainment areas. Measures must also ensure protection of areas that already meet NAAQS standards. Pursuant to the EPA's "Prevention of Significant Deterioration" (PSD) rule, the agency refused to certify plans that did not comply with this requirement. These legal opinions and regulations sparked major Congressional debate in 1976 and led to revised language in the 1977 amendments.

3.2 Legislative and Administrative Adjustments in the 1977 Amendments

The 1977 Clean Air Act amendments acknowledged that compliance schedules for SIPs and mobile source emission controls were lagging. Congress responded by extending the deadlines for four of the six designated criteria pollutants. State compliance time frames for NO_x and carbon monoxide standards were extended to 1987. The amendments also codified EPA policies for preventing significant air quality deterioration in areas that were already meeting specified ambient air quality standards.

Compared to the limited number of advocacy groups participating in 1970 deliberations, political action groups emerged as significant players in 1977 revisions. Senator Howard Baker described this contrast as follows:

> *Every day in 1975 and 1976 when we considered and then conferred with the House on clean air amendments, our venues were packed with lobbyists and long lines were out doors. While our committee continued to have the kind of serious debate that dominated our earlier deliberations, the existence of an audience changed the tenor, if not the substance.*[39]

Baker noted further that joint House–Senate Conference Committee deliberations were "... dominated by auto industry advocates and others whose objective was to

[35] Ibid., at p. 256. Note, "Prevention of Significant Deterioration and its Routine Maintenance Exception: The Definition Of Routine, Past, Present, and Future," 33 *Vermont Law Review* 785–86 (2009).

[36] Ibid.

[37] *Fri v. Sierra Club*, 412 U.S. 541 (1973).

[38] 39 Fed. Reg. 42,514 (December 5, 1974). *See also* R. ShepMelnick, *Regulation and the Courts: The Case of the Clean Air Act* 71–75 (1983).

[39] Howard H. Baker, Jr., "Cleaning America's Air—Progress and Challenges," Remarks at Edmund S. Muskie Foundation Symposium, The University of Tennessee, Knoxville, March 9, 2005.

weaken the Clean Air Act."[40] A prominent corporate counsel described this contentiousness from a different perspective:

> *Industry felt it had no other choice but to attack virtually all regulations that were promulgated. On the other side, regulations were uniformly attacked by the citizens groups that argued that the regulations were not tough enough. These were not just blind attacks by both sides on the new legislation, but rather an attempt to obtain clear regulatory interpretations that were not forthcoming from the regulators.*

The amendments sought a political compromise between shorter-term attainment of air quality objectives and industry contentions that initial deadlines were not realistic. Provisions deferred deadlines for vehicle emission compliance to 1982.

The 1977 law also directed EPA to revise its PSD rules to provide increased latitude for state implementation programs. In announcing the revised rules in 1978, Administrator Douglas Costle highlighted their cost-effective streamlining of regulatory processes applicable to new sources. He pointed to their balance between safeguarding "important and widespread areas of clean air ..." and encouraging but not dictating state land-use decisions "... within a framework that supports the goal of clean air preservation."[41] An open letter from the Acting Director for Chicago-based Region V asked that industry, local and state governments partner to "... insure responsible growth, while protecting public health and property."[42]

The EPA's amended PSD standards shifted from an emphasis on air quality attainment strategies toward technical requirements applicable to new emission sources. This latter approach identifies the "BACT" for particular classes, such as coal-fired energy facilities, and then determines compliance based on whether those sources are functioning accordingly.

Consistent with the Supreme Court's *Union Electric* ruling, sources could not be directly exempted based on economic hardship or logistics in gaining and applying these technologies. A 1980 training manual states that that "[t]he basic goal of prevention of significant deterioration (PSD) is to ensure that air quality in clean air areas does not significantly deteriorate while maintaining a margin for future industrial growth."[43] In a same-year article for *EPA Journal*, Administrator Train opined that the strong progress resulting from the Act's technology-forcing approach outweighed disadvantages in increased costs, short-term inefficiencies in industrial processes, and delays in NAAQS compliance.[44]

[40] Ibid.

[41] EPA Announces New Rules on Industrial Growth in Clean Air Areas, EPA press release, June 13, 1978.

[42] Valdas V. Adankus, "An Open Letter to Elected Officials, Business Leaders, and Concerned Citizens," Acting Director, EPA Region 5, April 1978.

[43] EPA, *Prevention of Significant Deterioration Workshop Manual* (October 1980), at p. I-A-1. *See also* Craig N. Oren, "Prevention of Significant Deterioration: Control-Compelling Versus Site-Shifting," 74 *Iowa Law Review* 1 (October 1988).

[44] Russell E. Train, "EPA's Task," *EPA Journal* (November/December 1980).

3.3 The EPA in the 1980s: "New Federalism"

During the 1980s, Clean Air Act administration reflected national policies of "New Federalism" promoted by President Ronald Reagan. This approach emphasized downsizing the federal government and delegating more control to the states. The controversial 1981–1983 term of Administrator Ann Gorsuch illustrated this policy shift dramatically. The EPA's overall 1981 budget was reduced by nearly 30% (from $4.7 to $3 million). By early 1983, there were 20% fewer employees (11,400).[45]

A series of Congressionally mandated air pollution progress reports illustrated the administrative impacts of EPA's retrenchment.[46] For 1980–1981, the agency highlighted its streamlining of SIP reviews by tailoring the extent, reducing the steps and total processing time, and conducting state and federal reviews concurrently.[47] The 1982 update cited the success of this strategy in reducing backlog by 93%. Unsurprisingly, it considered these actions as key to improved relationships with states by reducing compliance uncertainty. The 1982 report promised further commitment to expediting SIP reviews and delegating more program management to state and local pollution control agencies.[48] In practice, this brought businesses into more direct interactions with state regulators rather than with EPA's Washington or regional offices. Administrator Gorsuch resigned on March 9, 1983. She referred to "... intense Congressional controversy about Administration policies"[49] in her public statement about this decision.

William Ruckelshaus returned for a second term as EPA Administrator from mid-1983 to January 1985. During that period, the agency integrated risk assessment and risk management into its decision-making processes.[50] On November 2, 1983, it directed states either to provide evidence of NAAQS compliance or commit to remedying deficiencies in their SIPs to avoid potential sanctions.[51] In June 1985, the agency announced "a more aggressive direct Federal regulatory program that involves utilization of all EPA authorities, not just the Clean Air Act ..." to reduce air toxics.[52]

Building on an approach initiated in earlier pilot programs, the EPA issued an Emissions Trading Policy Statement in 1986. This "bubble area" strategy allows sources within a designated area to exchange emission reductions attained through less costly processes with allowances for releasing pollutants from more expensive technologies. These "trades" would be allowed only if the total net emissions within

[45] *See* Phil Wisman, "EPA History, (1970–1985)," *EPA Journal* (November 1985).

[46] *See* Lois R. Ember, "EPA Administrators Deem Agency's First 25 Years Bumpy but Successful," *Chemical and Engineering News*, October 25, 1995; Wisman, "EPA History, 1970–1985," *supra* note 45.

[47] EPA, *Progress in the Prevention and Control of Air Pollution in 1980 and 1981, Annual Report of the Administrator of the Environmental Protection Agency to the Congress of the United States*. EPA, 1981.

[48] EPA, *Progress in the Prevention and Control of Air Pollution in 1982, Annual Report of the Administrator of the Environmental Protection Agency to the Congress of the United States*, at p. V-2. EPA, 1982.

[49] Ibid.

[50] *See* Ruckelshaus, "William D. Ruckelshaus: Oral History Interview," *supra* note 12.

[51] EPA, *Progress in the Prevention and Control of Air Pollution in 1983*, at pp. VI-1–VI-2 (February 1985).

[52] EPA, *Progress in the Prevention and Control of Air Pollution in 1985*, at p. I-6. EPA, 1987.

the bubble area would not exceed NAAQS. At the time, EPA had already approved or proposed approval for 50 SIP revisions in 29 states that incorporated this strategy.[53] However, an April 27, 1987 letter from Administrator Lee M. Thomas to 42 Governors warned that sanctions such as withholding funds for federally assisted projects could be invoked for SIP noncompliance with ozone and carbon monoxide standards.[54]

By the late 1980s, the EPA showed greater signs that it was moving toward improved relations with regulated industries. An internal task group proposed that its SIP procedures set limits on internal reviews, provide greater certainty in the decision process, and delegate greater authority to the regional offices.[55] It proposed implementing these recommendations in 1988 by establishing "completeness criteria" for regional office review. The agency would also allow certain delayed applications to be accepted in their present form.[56]

William K. Reilly, Administrator from 1989 to 1993, emphasized alternatives to costly adversarial approaches that "do not frustrate the fundamental objectives either of the regulated sector ..., interest groups, environmental organizations and others ...."[57] These included voluntary programs, incentives, and negotiated rulemaking as preferred methods for policy development. This approach guided agency negotiations with the oil, auto, and chemical industries prior to adoption of the 1990 Clean Air Act amendments. Reilly noted in a 1995 retrospective article in *Chemical Engineering News* that relations with the industry improved notably during his tenure.[58]

> *At first, the chemical industry was largely resistant to a lot of new regulation from the federal government in an area that, in some cases, had previously not been regulated at all. But after the debacle of the early Reagan EPA, industry's attitude shifted dramatically. Industry had come to realize EPA was necessary to reassure the public that what industry was doing was consistent with public interest.*[59]

In a same-year oral history interview, Reilly offered broader support for collaborative methods:

> *I think regulatory negotiations are extremely productive at getting a result that works for everybody—where people don't hold back their best ideas so they can litigate them later; where you have a regulation promulgated that will, in fact, be the regulation, is not contested or litigated, permitting the regulated sector to invest on the basis of it ....*[60]

[53] EPA, *Progress in the Prevention and Control of Air Pollution in 1986*, at p. VI-1. EPA, 1988.
[54] "EPA Outlines Actions to be Taken with States Not Meeting Clean Air Standards; Some Sanctions Mandatory," Press Release, April 7, 1987.
[55] EPA, *Progress in the Prevention and Control of Air Pollution in 1987*, at p. VI-7. EPA, 1989.
[56] EPA, *Progress in the Prevention and Control of Air Pollution in 1988*, at pp. VI-11–VI-12. EPA, 1990.
[57] EPA, "William K. Reilly: Oral History Interview." http://nepis.epa.gov/Exe/ZyPURL.cgi?Dockey=40000BXV.txt (accessed October 2011).
[58] Ibid.
[59] Lois R. Ember, "Introduction to Special Issue on EPA's First 25 Years," *Chemical and Engineering News*, October 25, 1995.
[60] Ibid.

However, beyond the promise of these dispute–resolution alternatives, Congress and the courts can be expected to intervene when EPA, industry, and state interests cannot be resolved through consensus processes.

3.4 Implementation Issues in the 1990 Clean Air Act Amendments

The 1990 amendments provided substantially increased authority for the EPA to regulate toxic air emissions. Title V authorized the establishment of a national permit program for 189 pollutants specified within the legislation. Congress directed the agency to establish new industry-by-industry standards for source emitters in these categories. Criteria would be based on "maximum available control technology" that was currently in use for each industry.[61] Implementation of this expanded regulatory strategy brought a wide range of smaller-scale businesses into federal jurisdiction.

A second set of requirements directed the EPA to assess remaining risks to public health and the environment by 2004. Additional emission control standards would be issued within this time frame if warranted by these studies.[62] Other provisions in the 1990 amendments addressed acid rain, urban air pollution, and mobile source emissions.

During the first phase of Title V implementation, the EPA was required to issue emission control standards for 25% of its identified pollutant classes. Regulations for an additional 25% were to be in place by November 1997, and all applicable rules were to be established by November 2000.[63] These required actions would be carried out at the federal level. However, states would be responsible for ground-level implementation and for financially supporting their allocation of Title V costs through permit fees. A 1991 report by the General Accounting Office (GAO) expressed concern with EPA's implementation capacity.[64] It noted particularly that budget requests for the 1991 and 1992 fiscal years "were 23 and 16 percent, respectively, of the funds it deemed necessary to implement its air toxics program."[65]

[61] *See generally* James C. Robinson and William S. Pease, "From Health-Based to Technology-Based Standards for Hazardous Air Pollutants," 81 *American Journal of Public Health* 1518 (1991). "The striking political feature of these amendments is the virtual unanimity with which the usually fractious participants in air toxics debates agreed to the final compromise. This convergence is a political response to the fundamental weakness of the original statute." Ibid.

[62] *See* Clean Air Amendments of 1990, Title III.

[63] Clean Air Act Amendments of 1990, Title III. *See* U.S. GAO, Report to the Chairman, Subcommittee on Oversight and Investigations, Committee on Energy and Commerce, House of Representatives, *Air Pollution: EPA's Strategy and Resources May Be Inadequate to Control Air Toxics* 3 GAO/RCED-91-143 (June 1991).

[64] Ibid., at p. 3.

[65] Ibid., at p. 3.

3.4.1 The EPA Continues Its Moves Toward Collaborative Implementation

The EPA continued to redefine its policy-development approach following passage of the 1990 Clean Air Act amendments. Requirements to promulgate source-specific administrative rules for 189 toxic pollutants provided strong impetus for integrating interest group participation into these processes. A 1990 internal task force found that approximately 50% of EPA's adopted rules took between 16 and 50 months to adopt, and that at least 25% involved more extended time frames.[66] It recommended that the agency encourage greater interaction with affected interests and streamlining of internal processes.[67]

In early 1991, Administrator William Reilly promised collegiality and cooperation with state and local officials, industry, labor, and environmental groups. Elements of this strategy included establishing "advisory committees, regular informal consultations, and a formal regulatory negotiation process."[68] A June 1991 GAO report expressed cautious optimism that early consensus-building efforts were "... a good first step toward expediting the issuance of air toxics regulation."[69] An Office of Air and Radiation (OAR) report to the Administrator on the first 2 years of 1990 amendments implementation noted that:

> *Not only representatives of industry and environmental groups but also state and local government and public health officials, labor, consumer and academic leaders, and many other stakeholders have actively participated in EPA's negotiated rulemaking process. No interested party has been excluded, and no constituency has gone unheard. The consensus-building process has enabled us to establish enforceable clean air standards, while accommodating the business plans of the most progressive companies and leveling the playing field for their competitors.*[70]

The agency's newly established Clean Air Act Advisory Committee (1992) also praised these initiatives:

> *EPA's approach to rulemaking—a collaborative negotiating process open to all parties—demonstrates a growing openness on the part of the Agency and a recognition that negotiation is often preferable to litigation. Flexible yet enforceable rules are emerging from this consensus–building process.*[71]

[66] EPA Clean Air Act Implementation Task Force, *Report to the Deputy Administrator*, at p. 4, July 1990. EPA, 1990.

[67] Ibid., at p. iii.

[68] William K. Reilly, "The New Clean Air Act: An Environmental Milestone," *EPA Journal*, January–February 1991, at pp. 3–4. *See also* Edward P. Weber and Anne M. Khademian, "From Agitation to Collaboration: Clearing the Air through Negotiation," 57 *Public Administration Review* 396 (1997); U.S. GAO, EPA's Strategy and Resources May Be Inadequate, supra note 63, at p. 13.

[69] U.S. GAO, *EPA's Strategy and Resources May be Inadequate*, 3, *supra* note 63, at p. 25.

[70] *Report of the Office of Air and Radiation to Administrator William K. Reilly, Implementing the 1990 Clean Act: The First Two Years*. EPA, 1992, p. 3; EPA Office of Air and Radiation, *Implementation Strategy for the Clean Air Act Amendments of 1990* (Update 1992), at p. 3.

[71] EPA Clean Air Act Advisory Committee, *The Clean Air Act of 1990: An Introductory Guide to Smart Implementation*, at p. 7 (1992).

Between 1990 and 1995, the OAR estimated that over 100 rulemaking processes had employed consultation and consensus-building techniques.[72] This included five formal negotiation processes: Small Non-Road Engines Emissions Controls, Reformulated Fuels, National Emissions Standards for Coke Oven Batteries, volatile organic compound emissions from the wood furniture manufacturing industry, and Architectural and Industrial Maintenance Coatings.[73] These processes contrast with earlier approaches where rules would be developed within the agency and then presented in completed draft form for review and comment by affected stakeholders. In that framework, affected industries are left only with the options of policy acceptance or legal challenge.

EPA established a National Environmental Performance Partnership System (NEPPS) in 1995 to encourage better relationships with the states. Under this program, the EPA and states negotiate Performance Partnership Agreements that establish priorities so that both work jointly toward these objectives. Forty-five states were participating in this program by 1998. The GAO reported concerns in 1999 with the EPA's consistency of oversight across regions, detailed intervention into state programs, limited technical support, and limited consultation with states on key decisions affecting them.[74] This report also noted that the Washington office sometimes treats NEPPS negotiations as a regional-state matters, and "… do not view themselves as 'signatories' to the process."[75]

3.4.2 Collaborative Implementation of the Title V Permitting Program

The 1990 amendments established an intertwined regulatory program for the 189 toxic pollutants listed within its provisions. Title V created a federal permitting system for these source emitters; however, states would be required to administer and financially support this program. EPA estimated that nearly 17,000 major industrial sources were within this scope through 2006[76] (Figure 3.3).

Agency policies in the mid-1990s reflected President Clinton and Vice President Gore's initiative to "reinvent" government. A National Performance Review report (1995) on environmental regulation recommended that an adversarial "command and control" approach be complemented by more collaborative processes:

> *We have learned that the adversarial approach that has often characterized our environmental system precludes opportunities for creative solutions that a more collaborative system might encourage. When decision-making is shared, people can bridge differences, find common ground, and identify new solutions. To reinvent environmental protection, we must first build trust among traditional adversaries.*[77]

[72] U.S. GAO, *Clean Air Rulemaking: Tracking System Would Help Measure Progress of Streamlining Initiatives* 18 (March 1995).

[73] Ibid.

[74] U.S. GAO, *Environmental Protection: Collaborative EPA-State Effort Needed to Improve New Performance Partnership System* 2 (June 1999).

[75] U.S. GAO, *Environmental Protection: Collaborative EPA-State Effort Needed*, at p. 48.

[76] Congressional Research Service, *Clean Air Permitting: Status of Implementation and Issues*, at p. CRS-2 (2006).

[77] President Bill Clinton and Vice President Al Gore, *Reinventing Environmental Regulation*, at p. 3 (March 16, 1995).

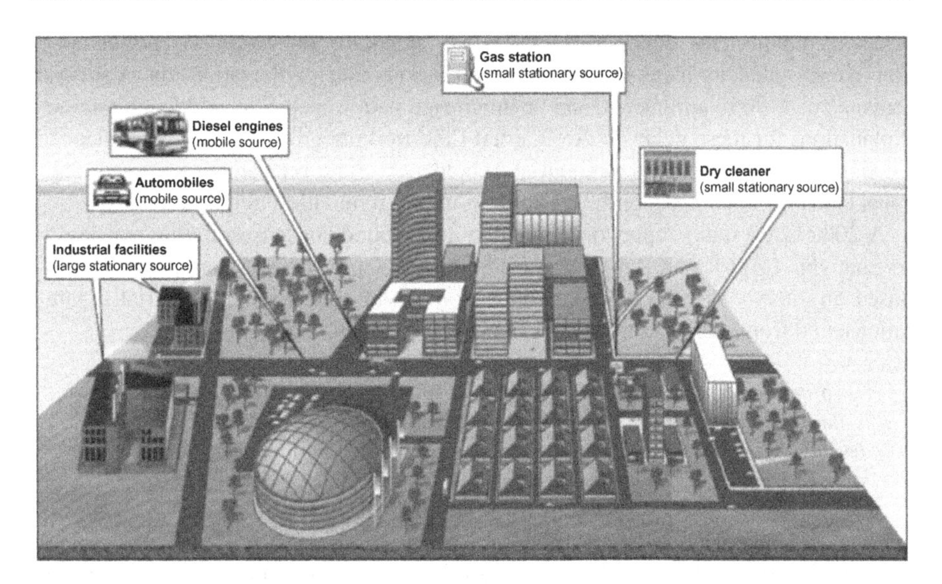

Figure 3.3 Common sources of air toxic emissions.
Source: GAO, EPA Should Improve the Management of its Air Toxics Program, at p. 9 (July 2006).

Administrator Carol Browner established an Office of Reinvention to coordinate agency efforts.

While consensus-building approaches may have diverted resolution from judicial outcomes, they also raised expectations from participants. A GAO report (1997) noted stakeholder concerns over "the large number of complex and demanding initiatives ... as well as confusion over the underlying purpose of some of the agency's major initiatives."[78] Regulators, industry, and environmental interest participants expressed frustration where lack of unanimous consent kept negotiations from moving forward.[79] In such cases, an overemphasis on avoiding litigation could have led to delays in issuing regulations. A 1998 internal audit recommended that the agency clarify expectations, process, and expected outcomes in its relations with stakeholders.[80] The EPA continues its proactive measures to bringing industry, state, environmental, and other defined interests into policy-development negotiations.

3.5 The EPA Confronts Twenty-First Century Challenges

By early 2000, the EPA had met approximately three-fourths of its 538 required actions under the first six titles of the 1990 amendments. This included decisions on 198 matters that had passed beyond their statutory deadlines. A portion of these

[78] U.S. GAO, *Challenges Facing EPA's Efforts to Reinvent Environmental Regulation*, at p. 6 (July 1997).
[79] Ibid., at p. 9.
[80] EPA Office of Inspector General, *Audit Report, The Effectiveness and Efficiency of EPA's Air Program*, pp. 29–30 (1998).

delayed commitments concerned setting new standards for hazardous pollutants.[81] The agency cited its increased emphasis on stakeholder participation, the workload demands of 1990 requirements, and unanticipated policy, technical, and legal issues as explanations for these delays.[82] An internal EPA working group (2000) recommended creating a database for key national stakeholders.[83] State regulators expressed additional concerns about the significant burdens imposed by Title V requirements.[84]

A 2000 GAO status report on implementation found consensus among stakeholders that the 1990 Clean Air Act amendments had positive environmental effects. Based on interviews with representatives from environmental and industrial organizations, and from state and local governments,[85] it found that:

> *... [o]ne of the overarching issues affecting implementation cited by stakeholders is the tension between allowing states and sources of pollution the flexibility to develop their own approaches for achieving air quality improvements and using a more prescriptive "command and control" approach.*[86]

Interviewees criticized the lack of clear specifications in the statute and regulations and recommended greater flexibility for states to determine their strategies for achieving air quality improvements. There was added consensus that states and local governments lacked sufficient resources for implementation and enforcement.[87]

An internal report by EPA's Office of Environmental Policy Innovation (2001) reviewed formal stakeholder evaluations from public involvement processes. Its analysis identified the following as key elements affecting success: trust, credible data and technical assistance, linkage to economic and social concerns, staff understanding of consensus-seeking processes, effective interactions with experts, and recognition of barriers to participation.[88] The agency initiated a site-based voluntary performance tracking program in cooperation with state environmental agencies. It also created a Compliance Assistance Advisory Committee within its National Advisory Council on Environmental Policy and Technology. Membership included representatives from businesses and trade associations; tribal, state, and local governments; community and environmental groups; and other federal agencies.[89]

[81] U.S. GAO, *Status of Implementation and Issues of the Clean Air Act Amendments of 1990*, at p. 10 (April 2000).
[82] Ibid., at p. 11.
[83] EPA, *Engaging the American People, A Review of EPA's Public Participation Policy and Regulations with Recommendations for Action*, at p. 20 (2000).
[84] U.S. GAO, *Air Pollution: Status of Implementation and Issues of the Clean Air Act Amendments of 1990*, at pp. 6, 44 (April 2000).
[85] EPA Office of Inspector General, Audit Report, *The Effectiveness and Efficiency of EPA's Air Program*, footnote 11, at pp. 12–14.
[86] EPA Office of Inspector General, Audit Report, *The Effectiveness and Efficiency of EPA's Air Program*, footnote 11, at p. 13.
[87] Ibid., at pp. 4–5 (2000).
[88] EPA Office of Policy, Economics, and Innovation, *Stakeholder Involvement & Public Participation at the U.S. EPA: Lessons Learned, Barriers, & Innovative Approaches*, at pp. 1–2, 4–7 (January 2001).
[89] EPA Office of Compliance Assistance, *Compliance Assistance Activity Plan Fiscal Year 2001*, at p. 3 (2001).

A 2003 GAO survey of state environmental officials noted agreement that industry will have greater flexibility to modify existing facilities. More than half agreed that industry will have greater flexibility to modify existing facilities.[90] One observer noted that: "[t]he incorporation of all applicable requirements into one document gives the State, sources, and the public a clear picture of what is required of a source to maintain compliance with State and Federal air laws."[91] A subsequent GAO report (2004) also found consensus that deregulation for new sources would increase flexibility and certainty. However, a comparable number of respondents correlated that finding to increased area pollution. The agency's response contrasted these concerns with its assessment that ambient air quality would not be affected.[92]

In 2004, the National Research Council (NRC) issued a comprehensive report on *Air Quality Management in the United States.* Its five major recommendations were to: (1) strengthen scientific and technical capacity; (2) expand national and multi-state control strategies; (3) transform the SIP process; (4) develop an integrated program for criteria and hazardous pollutants; and (5) enhance ecosystems and public welfare.[93] For SIPs, it recommended shifting to an "integrated multipollutant air quality management plan and reform process to focus on tracking results using periodic reviews, encouraging innovative strategies, and retaining conformity and federal oversight."[94]

The EPA's Clean Air Act Advisory Committee responded by creating an Air Quality Management Work Group to address the five areas recommended by the NRC report.[95] It proposed convening scientists with affected stakeholders to develop a "framework for accountability" to strengthen air quality monitoring and reporting methods.[96] The Work Group also recommended streamlining SIP review by allowing states to revise noncompliant implementation plans without the EPA making a formal announcement prior to its remand.[97]

Title V regulation continues as a major implementation concern for the EPA and states. Table 3.1 reflects the scope of required EPA actions to designate and issue Maximum Achievable Control Technologies (MACT) criteria. By April 2005, the EPA had completed 404 of its mandated 452 actions required by the 1990 amendments. However, over three-fourths of its requirements with statutory deadlines were accomplished late. Within those 256 postdeadline actions, 96 were completed more than 2 years after their scheduled completion dates.[98]

[90] U.S. GAO, *Clean Air Act: EPA Should Improve the Management of its Air Toxics Program*, at p. 5 (June 2006).

[91] Ibid.

[92] EPA, *Clean Air Act: Key Stakeholders' Views on Revisions to the New Source Review Program*, at p. 13 (February 2004).

[93] National Research Council of the National Academy of Sciences, *Air Quality Management in the United States*, at p. 5 (2004).

[94] Ibid., at p. 14.

[95] EPA Air Quality Management Work Group, *Recommendations to the Clean Air Act Advisory Committee* (2005).

[96] Ibid., at pp. 16–17.

[97] Ibid., at pp. 26–7.

[98] U.S. GAO, *Clean Air Act: EPA Has Completed Most of the Actions Required by the 1990 Amendments, but Many Were Completed Late*, at pp. 3–4 (May 2005).

Table 3.1 Number of Air Toxics Actions Required Under the 1990 Clean Air Act Amendments

Air Toxics Category	Number of Actions Required
Major stationary sources regulated by MACT standards	158
Eight-year residual risk reviews for MACT standards	96
Eight-year technology reviews for MACT standards	96
Standards for small stationary sources	70
Mobile sources	2
Other (reports, studies, etc.)	31
Total	**453**

Source: United States Governmental Accounting Office, Clean Air Act, EPA Should Improve Management of its Air Toxics Program (2006), at p. 14. GAO analysis of EPA documents.

The Governmental Accountability Office (2006) criticized the EPA for lacking a comprehensive implementation strategy for air toxics regulation. Stakeholders expressed concern that the program suffered from lack of available data to assess effectiveness, inadequate funding, and its low priority relative to other clean air programs.[99] The GAO also reported that Title V had changed company practices by shifting more environmental responsibilities to operations personnel and to executives. Industry representatives voiced their discomfort that regulatory uncertainty seemed to be a negative effect of new source rules. An added perception was that increasing compliance costs netted minimal environmental improvements.[100] Senior EPA officials responded critically that external stakeholders often define the air toxics program agenda by litigating when the agency misses its deadlines.[101]

Administrator Christie Todd Whitman announced an alternative collaborative strategy in 2006 developed by senior leaders that were convened as an Innovation Action.[102] While this council praised the success of the NEPPS and State/EPA Innovation Agreements, it recommended that agencies should de-emphasize regulation as the prevailing implementation method:

> *Government should create more financial incentives for strong environmental performance. These incentives should be used in regulatory and nonregulatory programs, and they should take many different forms—ranging from trading programs that provide flexible, cost-effective compliance options for industrial facilities to liability provisions that reduce costs for safer, cleaner operations.*[103]

[99] Ibid., at pp. 23–4.

[100] U.S. GAO, *Air Pollution: Status of Implementation and Issues of the Clean Air Act Amendments of 1990, supra* note 81, at p. 46.

[101] U.S. GAO, *Clean Air Act: EPA Should Improve the Management of its Air Toxics Program*, at p. 5 (June 2006).

[102] U.S. EPA, *Innovating for Environmental Results: A Strategy to Guide the Next Generation at EPA* (April 2002).

[103] Ibid., at pp. 4–5.

Administrator Whitman emphasized an internal top–down commitment to patiently nurturing relationships with major partners to reach consensus toward "measurable, affordable progress." She encouraged staff to "... begin to see their job as environmental problem solvers—helping to develop new tools and creatively applying them to solve specific environmental problems."[104]

The EPA continues to encourage collaborative processes among federal, state, and local air pollution control officials.[105] The 2010 Agency Financial Report highlights these relationships within the federal government; with states, localities, and tribes; with business and industry; and with nonprofit organizations and educational institutions: "EPA understands that government alone cannot begin to address all of the nation's environmental challenges."[106]

3.6 EPA Budget and Staff Resources

The EPA began operations in 1970 with 4,000 employees and an overall $1 billion budget allocated primarily to programs transferred from other agencies.[107] By FY 1975, more than 10,000 staff members were managing a $6.9 billion budget. After sharp decreases in 1977, there were approximately 12,000 personnel managing $5.4 billion in FY 1979.[108] Senator Muskie and others voiced serious concern during the 1980s over waning support for air quality and other environmental programs during that time.[109] EPA budget trends changed significantly with passage of the 1990 Clean Air Act amendments. Its budget allocation for 1991 reached $6.1 million.[110]

According to a 1994 GAO report, the EPA's air quality programs were at $520 million. This was about 8% of the agency's overall budget.[111] However, this amount was $25.4 million less than was requested by the agency.[112] EPA officials reported that these budget cuts would delay the statutory schedule for establishing industry-specific emission standards due in 1992 and 1994, and that it was suspending work on standards scheduled for completion by 1997.[113] In FY 1996, the EPA's OAR spent $175.4 million administering Clean Air Act programs.[114]

[104] Ibid.

[105] EPA, *2006–2011 EPA Strategic Plan Charting Our Course* 14 (October 2006).

[106] EPA, *Fiscal Year 2010 Agency Financial Report*, at p. 3 (2010).

[107] EPA, "EPA Budget and Spending." http://www.epa.gov/planandbudget/budget.html (accessed October 2011).

[108] Ibid.

[109] *See* Edmond S. Muskie, "NEPA to CERCLA, The Clean Air Act: A Commitment to Public Health," *The Environmental Forum* (January/February 1990). http://www.cleanairtrust.org/nepa2cercla.html (accessed October 2011).

[110] *See* EPA, "EPA's Budget and Spending." http://www.epa.gov/planandbudget/budget.html (accessed October 2011).

[111] U.S. GAO, *Air Pollution: Reduction in EPA's Air Quality Budget* 8 (November 1994).

[112] Ibid., at p. 9.

[113] Ibid., at p. 19.

[114] EPA Office of Inspector General, Audit Report, *The Effectiveness and Efficiency of EPA's Air Program* 3 Report No. E1KAE4-05-0246-8100057 (1998).

EPA budget levels have been relatively consistent since 2000. It exceeded $8 million in 2005. In that year, approximately 15% (2,644) of the agency's 17,500 employees were administering Clean Air and Climate Change programs.[115] For FY 2010, Clean Air and Global Climate Change programs constituted approximately 10% of the agency's overall $10.48 million budget.[116] While these resources increased consistently over 35 years, variations correspond to Congressional and Executive commitments to environmental programs.

3.7 EPA's Present Structure and Functions

The EPA's present organizational structure retains the basic relationships from when the agency was established. There is a continuing direct line of authority between the Administrator and 10 regional offices. The combined Air and Radiation Office is at a parallel level with other environmental programs. Its programs focus on air quality regulations and technical policies on issues such as industrial and vehicle pollution, acid rain, ozone depletion, and climate change.[117]

The changing role for regional offices is reflected in a comparison of the 1983 and 2000 editions of its *Organization and Functions Manual*. The 1983 version states that generally that these offices "develop, propose, and implement an approved Regional program for comprehensive and integrated environmental protection activities."[118] The manual published in 2000 offers a more comprehensive role in enforcement and compliance:

> *Regional Administrators are responsible for developing, proposing, and implementing regional programs for comprehensive and integrated environmental protection activities; conducting effective regional enforcement and compliance programs; translating technical program direction and evaluation provided by various Assistant Administrators, Associate Administrators and Heads of Headquarters Staff Offices into effective operating programs at the regional level, and assuring that such programs are executed efficiently;* **exercising approval authority for proposed State standards and implementation plans; and providing overall and specific evaluations of regional programs**[119] *[emphasis added].*

This comparison reflects an expanding regional role from facilitating communication between states and headquarters to frontline exercise of national authority. The Washington-based Office of Regional Operations provides advocacy and ombudsman functions. This includes enhancing regional participation in agency decision-making and ensuring that policy directives are conveyed effectively.[120]

[115] EPA, *FY 2006 Annual Plan*, at p. G/O-1 (2006).
[116] EPA, *FY EPA 2010 Budget in Brief*, at p. 5 (2009).
[117] *See* "EPA Office of Air and Radiation." http://www.epa.gov/air/index.html (accessed July 2011).
[118] EPA, "Region Functional Statements," United States Environmental Protection Agency 21 (September 2007).
[119] EPA, "Region Functional Statements," *Organization and Functions Manual* (November 2000).
[120] EPA Office of Regional Operations, "What We Do." http://www.epa.gov/aboutepa/ao.html#oro (accessed October 2011).

Overall, the EPA remains committed to resolving matters short of litigation. The FY 2010 report describes the agency's operating strategy as follows:

> *How EPA Works: Collaborating With Partners and Stakeholders*
>
> *Addressing today's complex environmental issues requires greater transparency and cooperative action; establishing and enhancing working partnerships; and combining EPA's resources with those of other federal agencies and state, local, and tribal partners. EPA understands that governmental one cannot begin to address all of the nation's environmental challenges. The Agency also works with business and industry, non-profit organizations, environmental groups, and educational institutions in a wide variety of collaborative efforts.*[121]

3.8 EPA's Continuing Role: Balancing Consensus Initiatives with Legal Mandates

The EPA faces continuing challenges to ensure active stakeholder participation and respond to legal mandates regarding criteria for NAAQS and toxic pollutant regulations. While consensus processes have shown effectiveness in policy-development and regulatory issues, litigation retains its role as a means to resolve ongoing concerns. Conflict over climate change causes, impacts, and potential responses may continue as an EPA priority in its fifth decade.

Environmentalists initiated a challenge to the EPA's refusal to designate greenhouse gases (GHGs) in 1999. The EPA's General Counsel supported this position, but the agency did not take action at the time. Four years later, under the administration of George W. Bush, the agency denied this petition. In 2007, the Supreme Court in *Massachusetts v. EPA*[122] supported the state's contention that the agency must make a determination whether GHGs are to be classified as NAAQS pollutants. It found that the Clean Air Act's broad definition of "air pollution" could incorporate GHGs. The Court mandated that the EPA make an official finding but did not require the agency to initiate a regulatory strategy or other action.

In April 2009, the EPA declared its intention to find that GHG emissions from new vehicles would endanger public health and welfare because they contribute to global climate change.[123] This proposal stimulated significant controversy and challenges from industrial and other sectors. These challenges hit the core of scientific legitimacy concerning the causal link between these emissions and climate change impacts. The EPA issued a final Endangerment Finding in December 2009 that designated these combined gases as criteria pollutants subject to NAAQS.

[121] EPA, *Fiscal Year 2010*, Agency Financial Report, Section 1, at p. 3 (2010).

[122] 549 U.S. 497 (2007).

[123] Proposed Rule, 74 Fed. Reg. 18, 886 (April 24, 2009). *See also* Patricia Ross McCubbin, "Proposed Endangerment and Cause or Contribute Findings for Greenhouse Gases Under Section 202(a) of the Clean Air Act," 33 *Southern Illinois University Law Journal* 437, 438–39 (2009).

Industries and other stakeholders continue their active advocacy on appropriate national climate change policies. One prevalent strategy, referred to as "cap and trade," would extend the "bubble concept" to the national level by setting overall emission limits and allowing industries and other key sources that exceed permissible levels to "purchase" pollution rights from sources that are within applicable standards. Industry, government, and other sectors have expressed grave concerns with the scope and feasibility of this implementation strategy.

In June 2011, Administrator Lisa P. Jackson issued an agency policy statement on climate change adaptation and development of a response plan to be completed by June 2012. This announcement presages that science will continue to be invoked and interlinked in the evolution of federal policy development:

> *Scientific evidence demonstrates that the climate is changing in unprecedented ways. These changes can pose significant challenges to the EPA's ability to fulfill its mission. The EPA must therefore adapt to climate change if it is to continue fulfilling its statutory, regulatory and programmatic requirements, chief among these protection of human health and the environment. Adaptation will require that the EPA anticipate and plan for future changes in climate and incorporate considerations of climate change into many of its programs, policies, rules and operations to ensure they are effective under future climatic conditions. Through climate-adaptation planning, the EPA will also contribute to the federal government's leadership role in promoting sustainability and in pursuing the vision of a resilient, healthy and prosperous nation in the face of a changing climate.*[124]

[124] U.S. Environmental Protection Agency, Lisa P. Jackson, Administrator, "Policy Statement on Climate Change Adaptation," June 2, 2011.

4 State Implementation Planning for Clean Air

The 1970 Clean Air Act amendments placed unprecedented implementation demands on state governments. The EPA's initial regulations in 1971 required each to assign a lead entity to develop a timely implementation plan for the six designated NAAQS pollutants. State program responsibilities would include monitoring pollution sources (mobile and stationary), and determining regulatory strategies based on federally compliant implementation plans. They would also be positioned as frontline regulators for industries, motor vehicles, and other identified pollution sources. The 1970 law presented immediate challenges to staff and finance these programs beyond the limited assistance funds authorized by Congress.[1] The 1990 amendments added more unprecedented responsibilities to administer Title V federal permits for the 189 toxic chemicals it specified. This chapter reviews the structural responses, funding, regulatory strategies, and collaborative measures in state responses since the 1970 Clean Air Act amendments. It incorporates survey responses from state environmental officials as part of a study conducted by the Environmental Research Institute at the University of California, Riverside, and supported by a grant from the Andrew J. Mellon Foundation.

As enacted, the 1970 Clean Air Act amendments required states to submit implementation plans to EPA for approval. States could attain these levels by regulating emission sources or indirectly through transportation planning and other means. Their laws could also require "owners or operators of stationary sources to install, maintain, and use emission monitoring devices and to report on the nature and amounts of emissions from such stationary sources ...." Further, states would be required to adopt motor vehicle inspection programs. The guidelines were clear as well that the agency reserved residual enforcement power when states fall short of their proposed implementation plans.[2]

Initially, the EPA estimated that 8,500 personnel would be required at the state and local levels to achieve overall compliance by 1975. Compared to an estimated 2,800 national total in 1969, there were approximately 6,200 employees in state and local air pollution control programs by late 1973.[3] Agency rules in April 1971 established an optimistic schedule that would attain universal NAAQS compliance

[1] Congress authorized $29 million in grants for state and local implementation of air quality programs in 1971, and $43 million for FY 1972. Environmental Protection Agency, *A Progress Report December 1970 to June 1972*, at p. 5. EPA, 1972.

[2] EPA, *Environmental Protection Agency: A Progress Report*, December 1970–June 1972. Washington, D.C., 1972, pp.1–2.

[3] Ibid., at p. 149.

An Interactive History of the Clean Air Act. DOI: 10.1016/B978-0-12-416035-4.00004-7

by 1975. The EPA rejected all but 13 of the original state submissions and proposed that the remaining plans be adopted via federal authority as proposed by the agency.[4] A January 1974 compilation identified 16 fully approved SIPs and 5 additional plans with approved regulatory elements. There were 19 plans that included sections proposed by EPA, with the remainder awaiting EPA action.[5]

4.1 Evolving Program Structures in State Air Quality Programs

Prior to 1970, state air programs were limited primarily to monitoring, research, and local assistance. The examples in this chapter track the evolution of these programs as well as those established as direct responses to national legislation. These administrative responses can be classified into four main approaches.[6] Some states continued their delegation of air quality planning to existing public health agencies. A more prevalent response adopted by 19 states consolidated air pollution control with other environmental programs as a "mini-EPA" umbrella agency. Other states formed multifunction "super-agencies" combining environmental programs with other functional areas; for example, North Carolina's Department of Natural Resources and Economic Development.[7]

A fourth approach delegated authority to various decentralized boards and commissions.[8] In Illinois, the Illinois Environmental Protection Act of 1970 split policy, enforcement, and research functions into three agencies. The Pollution Control Board develops regulations, its EPA enforces them, and the Department of Natural Resources Institute for Environmental Quality provides research support.[9] The Pollution Control Board has dual functions as a quasi-judicial "science court" to rule on environmental cases.[10]

A 1974 survey of states, territories, and the District of Columbia found that 29 jurisdictions had consolidated air, water, and other environmental programs into a single

[4] EPA, *EPA Progress Report December 1970 to June 1972*, at p. 2 (1972).

[5] EPA, *Legal Compilation—Supplement II*, at p. 134 (1974).

[6] An EPA guidance manual (1974) presented four modes for state implementation: full preparation at the state level where there is no local capacity; state coordination with local and regional agencies; state-local cooperative agreements; and plans coordinated at the local-regional level but implemented by the state. EPA, *Guidelines for Air Quality Maintenance Planning and Analysis, Volume 2: Plan Preparation*, at pp. II-13–15. (1974).

[7] North Carolina Department of Environment and Natural Resources, "A Short History of the Department of Environment and Natural Resources." http://portal.ncdenr.org/web/guest/history-of-denr (accessed October 2011).

[8] Evan J. Ringquist, *Environmental Protection at the State Level*, at pp. 38–9 (M.E. Sharpe, 1993).

[9] Illinois Environmental Protection Agency, "History of the Illinois EPA." http://www.epa.state.il.us/about/history.html (accessed October 2011).

[10] Illinois Pollution Control Board, "Citizen's Guide to the Illinois Pollution Control Board—The Board and the Act." http://www.ipcb.state.il.us/AboutTheBoard/CitizensGuidetotheBoard.asp?Section=Act (accessed October 2011).

agency; 20 jurisdictions included air pollution programs within their health agencies; and 6 states implemented programs through agencies dedicated to air pollution control.[11] From the mid-1970s through 1995, 19 states transitioned to umbrella agencies that include water quality, toxic waste, and other environmental programs. Table 4.1 reflects survey responses from 20 state programs.

California's program evolved through decades of responses to pollution crises. The Air Pollution Control Act of 1947 enabled each county to establish an Air Pollution Control District. In 1967, its Legislature established the Air Resources Board (ARB) by consolidating responsibilities of the Bureau of Air Sanitation (established in 1955) and the Motor Vehicle Pollution Control Board (established in 1960). The following survey response describes some of these transitions in California's air quality management programs:

> *While some local air districts predate the federal Act, many county-level districts were consolidated after 1970 to better address common air quality concerns. In 1976, for example, the South Coast Air Quality Management District (SCAQMD) replaced a voluntary association of air pollution control districts in the Los Angeles region consisting of Los Angeles, Orange, Riverside, and San Bernardino counties ....*
>
> *The Bureau of Automotive Repair (BAR) and the California Department of Pesticide Regulation (DPR) also have responsibility for emission reductions to meet California's air quality goals. BAR's Smog Check Program was created in 1984 to reduce automobile emissions in specifically designated areas that failed to meet federal and state clean air standards.*[12]

This response reflects a strategy focused on strengthening intrastate regional structures, and controlling mobile sources by regulating smog emissions.

Texas initiatives date back to 1956 when its Department of Health, Division of Occupational Health and Radiation Control began a statewide air sampling program. Legislation in 1965 authorized creation of an Air Control Board to safeguard air quality through civil enforcement powers.[13]

> *The initial implementation of the Clean Air Act of 1970 was through Chapter 382 of the Texas Health and Safety Code, also known as the Texas Clean Air Act, which was originally published in 1965 .... The first State Implementation Plan to improve Texas air quality was published in 1972. The organization responsible for implementing the CAA has changed significantly over the years from a service within the Texas Health Department, to an independent Air Control Board, to a portion of a Natural Resource Conservation Commission.*[14]

[11] EPA, *1974 Legal Supplement*, Suppl. II, Vol. 1 (Air) (January 1974), at p. 147.

[12] See Appendix, at pp. 96–107. *See also* California Air Resources Board, "History of Air Resources Board." http://www.arb.ca.gov/knowzone/history.htm (accessed October 2011).

[13] Texas State Historical Association, "Texas Air Control Board." http://www.tshaonline.org/handbook/online/articles/mdtls (accessed August 2011). *See also* Environmental Institute of Houston, "The History of Air Quality." http://www.prtl.uhcl.edu/portal/page/portal/EIH/outreach/tfors/history (accessed October 2011).

[14] See Appendix, at pp. 134–137.

Table 4.1 Administrative Structures for State Air Quality Programs[a]

Initial Agency for Clean Air Act Implementation	Current Agency Administering Air Quality Programs (With Date of Transition)
Alabama Air Pollution Control Commission	Department of Environmental Management (early 1980s)
Alaska Department of Health and Social Services	Department of Environmental Conservation (mid-1970s)
Arizona Department of Health Services	Department of Environmental Quality (1986)
California Air Resources Board (1968)	California Department of Environmental Protection, Air Resources Board (1991)
Colorado Air Pollution Control Division	Colorado Department of Public Health and Environment–Air Quality Control Commission (1995)
Florida Department of Environmental Quality	(continuous)
Hawaii Department of Health	(continuous)
Idaho Department of Environmental Quality	(continuous)
Maryland Department of Air and R adiation Hygiene	Maryland Department of Environment (1987)
Michigan Department of Environmental Quality, Division of Air Quality	(continuous)
Mississippi Department of Environmental Quality	(continuous)
Missouri Department of Natural Resources Air Pollution Control Program (1974)	(continuous)
Clean Air Act in Montana	Montana Department of Environmental Quality (1995)
Nebraska Department of Environmental Quality	(continuous)
Clean Air Act of New York	N.Y. Department of Environmental Conservation (1974)
Oklahoma Department of Health	Oklahoma Department of Environmental Quality (1993)
Rhode Island Department of Health	R.I. Department of Environmental Management (1977)
Texas Health Department, Air Control Board	Natural Resource Conservation Commission (1993–2001); Texas Commission on Environmental Quality (2001)
Utah Air Quality Board	Utah Department of Environmental Quality, Division of Air Quality (1991)
Virginia Department of Air Pollution Control	Virginia Department of Environmental Quality (1993)

[a]See Appendix for state responses.

Texas consolidated administration of its environmental programs in 1993 by establishing a Natural Resources Conservation Commission, which was renamed as the Texas Commission on Environmental Quality in 2001.[15]

Arizona's approach reflects a prevalent response to incorporate air pollution control within an "umbrella" environmental agency. Prior to 1971, responsibilities rested within its Department of Health Services Division of Air Pollution Control. Following passage of the federal Clean Air Act, Arizona created an umbrella Department of Environmental Quality with an expanded air program to regulate stationary and motor vehicle sources. The following survey response traces this path:

> *Arizona policy for the control of air pollution began in 1962 with legislation authorizing the Arizona Dept. of Health Services (ADHS) to conduct air pollution studies and thereby qualify the state for federal grants. In 1967, ADHS was authorized to begin setting air quality standards, and to establish its Division of Air Pollution Control. After passage of the 1970 CAA, additional legislation enhanced the state's role in air quality management. Responsibility for regulation of major stationary sources of pollution (copper smelting, power generation, etc.) was assigned to ADHS, and enforcement procedures were established for vehicle emissions. During the 1970s, the emphasis shifted from stationary to vehicular sources.*
>
> *The first SIP was submitted in 1972. In 1986, the Arizona Dept. of Environmental Quality was authorized, and all environmental management was moved from ADHS to ADEQ. In 1987, the Omnibus Air Quality Act was passed, enhancing the role of ADEQ in securing improved air quality. Originally, the counties were responsible for most A/Q management, but during the 1980s all but Maricopa, Pima, and Pinal County ceded authority back to the state.*[16]

Oregon, Virginia, and Montana followed similar approaches by creating "mini-EPAs." Oregon established a State Air Pollution Authority in 1951 that merged its functions with the Sanitary Authority in 1959. In turn, these entities became part of the Department of Environmental Quality in 1969.[17] The Virginia Department of Environmental Quality assumed Clean Air Act responsibilities in 1993 from the former Department of Air Pollution Control.[18] Montana established a Department of Environmental Quality in 1995 to include air quality regulation with other environmental programs.[19]

Hawaii and Maryland built on activities that were previously based in their public health programs. Hawaii's expanded program "has ... responsibilities consisting of permitting, source monitoring, enforcement, ambient monitoring, laboratory support, and clerical support."

[15] Texas Commission on Environmental Quality, "TCEQ History." http://www.tceq.texas.gov/about/tceqhistory.html (accessed October 2011).

[16] See Appendix, at pp. 94–96.

[17] Oregon Department of Environmental Quality, "DEQ Timeline." http://www.deq.state.or.us/about/historytimeline-p1.htm (accessed October 2011).

[18] See Appendix, at pp. 138–139. *See generally* Elizabeth H. Haskell, "An Environmental History." http://www.deq.virginia.gov/history/haskell.html (accessed October 2011).

[19] See Appendix, at pp. 126–127.

Table 4.2 Survey Responses on Funding for State Air Quality Programs[a]

State	Initial Funding Sources	Current Funding Sources (Through 2002)
Alabama	State's general fund and EPA grand funds	EPA funds, non-Title V permitting fees and Title V emission fees
Alaska	Federal grant and state general fund monies	Permit fees
Arizona	State general fund, federal grants	State general fund, federal grants, Vehicle Emissions Inspection fund, Air Quality Fee Fund, and Air Permit Administration Fund
California	State general fund and grant funding from EPA	State general fund and state special funds: Motor Vehicle Account (funded by vehicle registration fees), Vehicle Inspection and Repair Fund (funded by smog check fees), Air Pollution Control Fund (funded by fees on vehicle manufacturers, stationary sources, and penalty assessments). Annual grant funding from U.S. EPA
Colorado	Federal funds	Fees from mobile and stationary sources
Florida	State general revenue and 105 Grants from EPA	State general revenue and 105 Air Pollution Control Grant from EPA
Hawaii	State general funds	Fees from the regulated air sources. State general funds and federal grants
Idaho	EPA 105 funds, state general funds, and special grant funds	When Title V program began, fees were added to the funding sources. EPA 103 funds were added when the PM2.5 monitoring program began
Maryland	State general funds and EPA grants	State general funds; EPA grants; various permits and license fees and penalties (special funds)
Michigan	Federal grant and state general funds. Budget subsidized by surveillance fees.	Since 1998, annual air quality fees based on emissions data have supplemented the federal grant and state general funds. Federal grants and state general funds
Mississippi	State general funds and federal grants	Title V fees
Montana	EPA grant funds and state general funds	Air quality permit fees, EPA grant funds and fees
Nebraska	State general funds, federal grants, local general funds	State general funds, federal grants, local general funds, and most recently Title V emission fees
New York	Federal funding	Title V and Federal funding
Oklahoma	EPA's grant funds and state appropriations	State appropriations, EPA grant funds, and annual operating fees and other fees for services rendered
Rhode Island	State level	Section 105 EPA grants and matching state appropriations, traditionally near the minimum requirement: Title V federal grants and Title V fees
Texas	General revenues with some federal grant funds	Partially funded through permitting fees (1984); added vehicle inspection fees (1985); emission fees and vehicle inspection to totally finance Clean Air Act implementation (1992)
Utah	State appropriations, federal grants, and fees	State appropriations, federal grants, and fees
Virginia	General funds and federal grants under Section 105 Clean Air Act	Funding from Title V permit program; Northern Virginia vehicle inspection/maintenance program fees

[a]See Appendix (compiled from the 2002–2003 survey responses).

Hawaii only has a state air program and has no local or county air agencies. Since Hawaii is an island state, one staff each is located on Kauai and Maui, and two are located on the island of Hawaii. The remaining staff is located on Oahu in the Honolulu office.[20]

Maryland transferred air quality control activities from its Department of Health and Mental Hygiene to the Air and Radiation Hygiene division in the newly formed Department of the Environment in 1987. Program functions have expanded from an initial focus on permitting and enforcement to include seven program areas: planning, mobile source control, monitoring and data management, asbestos, radiation, permitting, and compliance.[21]

4.2 Funding for State Programs

With respect to funding, states relied heavily on federal grants during their initial program years. Their challenges increased significantly in the early 1990s in order to administer Title V regulations for toxic pollutants. Administering these federal permits has required continual adjustments in fee schedules and appropriate penalties toward making each program self-supporting. Table 4.2 reflects survey responses showing the trend from primary reliance on EPA grants and limited state general revenue contributions toward self-supporting programs through permit fees and enforcement actions.

Alaska:

At [the] state level, ... air quality programs have been funded primarily by federal grant and state general fund monies. Since approximately 1995, permit fees have become a significant funding source.[22]

California:

The CARB [California Air Resources Board] has generally been financed from a mix of state general fund and state special funds. The special funds include the Motor Vehicle Account (funded by vehicle registration fees), the Vehicle Inspection and repair Fund (funded by smog check fees), and the Air Pollution Control Fund (funded by fees levied on vehicle manufacturers and stationary sources of pollution, and penalty assessments). The State also receives annual grant funding from ... EPA. The ration of funding from the General Fund and the Motor Vehicle Account has varied over the years depending upon the availability of those funds, and the emission source (stationary source activities are primarily funded by the General Fund while mobile source activities are primarily funded by the Motor Vehicle Account).[23]

[20] See Appendix, at pp. 113–114.
[21] See Appendix, at pp. 117–119.
[22] See Appendix, at pp. 93–94.
[23] See Appendix, at pp. 96–107.

Hawaii:

Initially and up to 1997, the air program was predominantly supported with state general funds. As a result of the federal Clean Air Act of 1990, the air program was restructured to be largely supported by fees from the regulated air sources. The air program still receives money from the state general funds and federal grants.[24]

Michigan:

This process was initially financed through federal grant[s] and state general funds. From the mid-1970s to the mid-1980s, the budget was also subsidized by surveillance fees. Since 1998, in response to federal CAA requirements, annual air quality fees based on emissions data have supplemented ... federal grant and state general funds.[25]

Texas:

Initially, all activities were funded through general revenues with some federal grant funds added later. In about 1984, Texas State Air Pollution Control [Board] activities became partially funded through permitting fees. In 1985 inspection fees were added to help finance the process. About 1,992 emission fees and vehicle inspection fees were used to totally finance implementation of the CAA [for the state's mobile source program].

The state budget for the organizations which implement air quality is contained in the biennial Texas State Appropriations Act, Clean Air Account ....[26]

Utah:

In the early years, implementation was split about evenly among state appropriations, federal grants, and fees. Since implementation of the Operating Permits Program adopted in the CAA amendments of 1990, ... [program funding] shifted to approximately 40% fees and 30% each from state appropriations and federal grants.[27]

These survey responses reflect similar trends in air quality program funding. Through the 1980s, fiscal support relied primarily on federal grants from the EPA. State general funds have supplemented these sources along with regulatory fees from mobile source programs and other special pollution control permits. Since 1990, programs have shifted by necessity toward self-support via Title V federal permit fees and state administration of vehicle emission regulations.

[24] See Appendix, at pp. 113–114.
[25] See Appendix, at pp. 119–122.
[26] See Appendix, at pp. 134–137.
[27] See Appendix, at pp. 137–138.

4.3 Regulatory Strategies in SIPs

Under the Clean Air Act, states are required to monitor air quality, regulate motor vehicle emissions, and ensure that stationary pollution sources do not significantly degrade air quality.[28] The following selected responses from the Mellon project survey illustrate a range of regulatory strategies to implement SIPs.

California Environmental Protection Agency

California has a long and successful history of leading the nation in implementing programs to improve air quality. The state's pioneering research on the causes, effects, and methods for control of air pollution provide a strong scientific foundation for these air quality programs. California adopted the nation's first auto tailpipe emission standards for hydrocarbons and carbon monoxide in 1966, followed by the first automobile NOx emission standards, [the] first use of three-way catalytic converters, limits on lead in gasoline, and vehicle on-board diagnostic requirements. Many of these initiatives have served as models for national programs.

Extensive air quality monitoring networks, emission inventories, and air quality modeling provide the technical foundation for California's programs and regulations. California's monitoring program collects real-time measurements of ambient pollutants at over 40 sites throughout the state. The data generated are used to define the nature and severity of pollution, assess risk, determine which areas are in attainment or non-attainment, identify pollution trends, support agricultural burn forecasting, and develop air models and emission inventories.

California is the only state authorized by the federal act to adopt its own motor vehicle emissions or fuel standards and leads the world in advancing new vehicle technology. California's Low Emission Vehicle (LEV) program, adopted in 1990, treats vehicles and fuels as a system required to meet gradually decreasing in-use emission limits and has helped stimulate development of lower-emitting vehicles and cleaner fuels. In 1998, ARB adopted LEV 2 standards to further reduce NOx and VOC emissions from cars and require light trucks to achieve automobile emission limits. The U.S. EPA adopted more stringent national vehicle emission limits based on LEV 2 in 1999.

California's stationary source control program is implemented by the local districts. Each district implements its own New Source Review and stationary source permitting programs. Because of the severity of the air pollution problem, many new and innovative stationary source technologies are developed and applied in California. To encourage statewide consistency, ARB has developed Best Available Control Technology (BACT) requirements for power plants, refineries, smelters, and other stationary sources. California facilities generally emit far less pollutants per facility than most other facilities in the nation.

Some provisions of the federal act (or U.S. EPA interpretation of such provisions) have not provided California with the flexibility to effectively pursue its own proven toxic air contaminant control strategies. Implementation of the toxics elements of the 1990 Amendments consumed extensive resources with little health benefit beyond the preexisting state program. California has had particular difficulty implementing

[28] Erich Birch, "Air Quality Regulation in the United States," *Business Law Today* (July/August 2007).

its risk-based air toxics program—even though the requirements are likely to be at least as stringent as national standards—because U.S. EPA required a "line by line" equivalency demonstration.

Business assistance programs, run by ARB and local districts, increase compliance with California's air quality requirements. ARB provides technical training courses that keep industry and district enforcement personnel up-to-date on new technology and regulatory changes. ARB also enforces statewide control measures, and oversees district enforcement programs for stationary sources of pollution to ensure that emission reduction benefits are achieved and that all businesses are on level playing field.[29]

Florida Department of Environmental Quality

The state's air quality management program is largely driven by the requirements of the federal Clean Air Act and associated APE regulations. The DEP Division of Air Resource Management has overall responsibility for keeping track of these requirements and responding to them in a timely manner. This is accomplished within the division by organizational subunits having responsibility for the performance or statewide coordination of all activities related to air quality monitoring, emissions monitoring, PSD management, emissions inventories, air quality modeling, state rulemaking and SIP development.

Six DEP district offices and eight DEP-approved local air pollution control agencies have day-to-day operational responsibility for many routine air program functions such as air monitor operations, non-PSD permit processing, compliance inspections, and complaint investigations. Division staff handles the more complex permitting activities, such as PSD permits and Title V permits for utility acid rain units, but mainly function in a planning and coordination role. The consolidation of all air program planning and coordination functions in a single organizational entity provides administrative efficacy.

Through various administrative mechanisms, the Division of Air Resource Management ensures that the DEP district offices and DEP-approved local air pollution control programs perform all state and federal air management functions as required. Examples of these administrative mechanisms are as follows:

General Coordination and Oversight—This includes routine e-mail and telephone communications among division, district and local air program offices; periodic meetings of division/district/local air program managers; coordination of EPA grant air planning agreement commitments between the division and six EPA-funded local air programs, exchange of monthly activity reports, and occasional program audits of district/local air program functions by division staff.

Conference Calls—To identify air permitting and compliance problems and promote statewide consistency in how they are handled, monthly conference calls are held among both the division/district/local air permitting engineers and division/district/local air compliance staff.

Guidance Memoranda—To ensure statewide consistency on matters of rule interpretation and the like, guidance memoranda are drafted as needed by the division; circulated among the districts/locals, as well as the regulated community, for review and comment; and published in final form on the Internet.

[29] See Appendix, at pp. 96–107.

Specific Operating Agreements—Through specific operating agreements between the division and each of the eight DEP-approved local air pollution control agencies, various air program responsibilities, including compliance inspections, enforcement, and certain kinds of air permitting, are delegated to the local air programs.

Local Program Contracts—Contracts between the division and each of the local air programs are used to transfer funds from the statewide Air Pollution Control Trust Fund to the local programs for statewide Air Pollution Control Trust Fund to the local programs for support of certain air monitoring and Title V permitting and compliance-related activities.

Statewide Database systems—Statewide database systems for storage and retrieval of air quality monitoring data, emissions inventory data, compliance inspection and test data, and permit tracking data, are maintained by the Division and used by all district and local air program offices.

Statewide Training—The Division coordinated a statewide air training program including sponsorship of air pollution training courses by outside providers; annual workshops for air monitoring, permitting, compliance and emissions inventory staff; training field staff on use of air database systems; and a statewide annual air program meeting for all air program staff.[30]

Hawaii Department of Health

Hawaii has a network of approximately 18 ambient air quality monitoring stations which provides to the air program an assessment of the air quality throughout the state. In regard to permitting, the sources are categorized into agricultural burning, minor, CAA Title V minor, CAA Title V major, and PSD major sources. The amount and type of information required for an air permit application for a new source or modification are dependent on the site; building dimensions and distances; air quality monitoring data; meteorological data; air modeling with/without surrounding sources; BACT completed, a request for public comments, a public meeting, or a public hearing may be initiated depending on source category or the community sensitivity to the source.[31]

Michigan Department of Environmental Quality, Division of Air Quality

The state of Michigan's New Source Review (NSR) program is administrated through a central office in Lansing. The Permit Section consists of four units. One unit handles administrative functions, such as scheduled public comment periods and managing the section's permit database. The other three units perform the permit function. Each unit handles specific source categories. This structure was implemented in the late 1980s. Prior to this structure, the two units handled permit review based on geographical area. The change in structure occurred because the geographic setup was too often causing inconsistency in permitting for the same source type.

Best available control technology (BACT) is implemented in two primary ways in Michigan. For major sources, BACT is implemented pursuant to the federal Prevention of Significant Deterioration (PSD) program. Michigan implements a delegated PSD program pursuant to 40 CFR 52.21. For minor sources, BACT is

[30] See Appendix, at pp. 109–113.

[31] See Appendix, at pp. 113–114.

only required for new sources of volatile organic compounds, and for sulfur dioxide from natural gas sweetening facilities. State rules approved as part of Michigan's SIP provide the regulatory authority. The rules, as codified in the Michigan Code of Regulation, are R336.1702 (a) and R336.1403 (4), respectively.

Michigan has a state air monitoring network and an emission inventory system. The emission inventory is updated annually, in conjunction with the fee program, for sources subject to Title V of the federal CAA. The air monitoring program complements the federally mandated program. The AQD has a unit that does all of the dispersion modeling required for permit reviews, emission inventory, and SIP work.

Michigan has implemented several processes in the Permit Section that have improved the NSR process. The most important of these was to establish a position to screen incoming permit applications for administrative completeness. Some 80 to 90 percent of permit applications submitted to the state are incomplete in some way. The screening process is very effective. It ensures that very incomplete applications are corrected before being assigned to a permit reviewer. It also provides an initial notification to the applicant that the application was received, and meets basic requirements for further processing.

Another process is the use of General Permits. Michigan has issued five General Permits. Each permit covers a source category as concrete crushers, ethylene oxide sterilizers, or small remediations. The General Permits are limited to small sources that meet certain requirements.[32]

Texas Commission on Environmental Quality [formerly the Texas Natural Resource Conservation Commission]

The SIP process does work and continues to improve, but it is also technically and functionally very complex and resource-intensive. Historically, most states' early attempts to control ozone focused on VOC reductions. To meet federal mandates, the commission has adopted numerous regulations controlling VOCs from marine vessel loading, vessel cleaning/degreasing, vent gases, surface coating and degreasing, and printing, among many other types of sources. These regulations have been successful in considerably decreasing VOCs and ambient ozone levels, but have not achieved compliance with the ozone standard.

The Air Permits division (about 200 positions) deals with permitting and authorization of all air emission sources within the state. Texas addresses every man-made source of air pollution, no matter how small. The permitting process applies BACT, at a minimum, to every source which applies for a permit. The permitting flow process is described on pages 201 through 210 of the Report to the Sunset Advisory Commission. The SIP process impacts both permitted and non-permitted sources of air emissions in Texas by making changes to the air quality rules. The Technical Analysis Division (about 300 positions) includes a Modeling and Data Analysis Section which conducts airshed modeling to demonstrate the reductions to specific source category emissions within the airshed resulting from changes to air rules to allow the area to meet national air quality standards. An Areas and Mobile Emissions Assessment Section develops criteria emission estimates for area and mobile sources using EPA-approved mobile source models and emission factors. The Industrial Emissions Assessment Section conducts an annual emissions inventory

[32] See Appendix, at pp. 119–121.

of point source emissions for SIP analysis and development. The Monitoring Operations Division (about 120 positions) deploys and maintains more than 120 air monitoring stations in Texas which monitor ambient levels of criteria pollutants and provide feedback to the state and federal governments on attainment status of national air quality standards.

In January 1997, the Commission proposed a program that, for the first time in Texas' air pollution control history, extended beyond the confines of the urbanized areas. The purpose of the regional strategy was to reduce ozone causing compounds in the eastern half of the state in order to help reduce background levels of ozone in both non-attainment areas as well as those areas close to noncompliance for the new 8-hour ozone standard. The commission recently adopted NOx regulations for power plants, cement kilns, and other major NOx sources in East and Central Texas to address regional ozone. This control strategy was introduced at the commission's initiative, not in response to any specific federal requirements.[33]

Survey responses indicated a range of regulatory approaches to implement state air quality programs. California's initiatives extend to various strategies to control mobile source emissions. Its stationary source program is delegated to air quality management districts. There are also voluntary control programs in cooperation with businesses. The air program within the Florida Department of Environmental Regulation uses a wide range of technical assistance and outreach techniques to forward program objectives. Michigan's Division of Air Quality within the Department of Environmental Quality divides its program into administrative and permitting functions. Staff for the Texas Commission on Environmental Quality conducts monitoring, technical analysis, assessment, and regulation of air emissions. This program extends beyond urban regions to other identified sources.

4.4 Collaborative Processes in State Air Quality Programs

State administrators referred to collaborative processes that they initiated or in which they participated. The survey responses below include agency-initiated policy dialogue and more formal participatory processes.

Alaska Department of Environmental Conservation

We collaborate extensively. Best examples are numerous external workgroups or project panels which are established in support of developing or redesigning an air quality or other environmental program. These panels are usually comprised of a very wide array of stakeholders from industry, local government, environmental groups and tribal organizations. In essence they create an informal regulation or Policy negotiations team; they are not given final decision making authority, but their recommendations usually carry a heavy preference for the agency decisions.[34]

[33] See Appendix, at pp. 134–137.
[34] See Appendix, at pp. 93–94.

California Environmental Protection Agency

The scope of California's air pollution problems requires effective collaboration with the regulated community. ARB's rule development process is public by design and provides opportunity to interact with industry both formally and informally. ARB hosts a variety of workshops for the regulated community and routinely consults with business and industry groups. In recent years, ARB and industry have collaborated to develop comments regarding implementation of the federal 8-hour ozone standard, and lobby U.S. EPA for flexibility in implementing the toxics permitting programs called for in the 1990 CAA Amendments.

Regulated Industry: Virtually every industry in the state—and many out-of-state industries—is affected by California's air quality program. Key industry stakeholders include agricultural interests, oil and gas operators, vehicle and engine manufacturers, and consumer product manufacturers. ARB actively encourages open dialogue with industry before and during the regulatory process, as well as continuing contact afterwards to ensure smooth implementation ....

Environmental and community organizations have been active in many of ARB's programs, including serving on advisory committees for the implementation of clean gasoline and ZEVs. These organizations are also crucial in public outreach efforts.

Governmental associations serve as a forum to improve communication and cooperation among governmental stakeholders, address intra- and interstate air quality concerns, and share technical expertise.[35]

Florida Department of Environmental Quality

The major industry-based organizations that assist the division in our air quality efforts include: Electric Coordinating Group, Pulp and Paper Association, and the Chemical Manufacturers Association. Other groups include the American Lung Association, and numerous legal groups that represent other industries such as the municipal waste combustors, sugar industry, etc.

We usually collaborate with the regulated industry prior to rulemaking and legislative processes and it has been an effective process.[36]

Missouri Department of Natural Resources Air Pollution Control Program

Involving the public in the process of making air quality rules ... [is intended to] to create fair, effective regulations that have broad support. In 1999, DNR continued its commitment to public participation by convening workgroups to help develop air regulations. A workgroup brings industry and the public together with government agencies to share concerns and exchange ideas while developing regulations.

The department also worked with leaders from industry, environmental organizations and local government to improve air quality in the Kansas City area. The department participated as a member of the Mid-America Regional Council, a metropolitan planning organization, in the development of an air quality improvement plan for the Kansas City ozone maintenance area .... In June 2000, DNR

[35] See Appendix, at pp. 96–107.
[36] See Appendix, at pp. 109–113.

participated in the Kansas City Fuels Summit. Discussion focused on determining a motor vehicle fuel strategy to improve air quality in the Kansas City ozone maintenance area.[37]

Oklahoma Department of Environmental Quality

As part of the rulemaking process, the Air Quality Division of DEQ often establishes workgroups comprised of industry representatives, division staff, and the interested public. As an example, the agricultural community plays a significant role in Oklahoma. While in the process of revising out regulation dealing with the Control of Emissions of Grain, Feed, and Seed Operations, workgroups with significant industry representation proved to be the most expeditious ways to reach common goals in the rulemaking process.[38]

These state collaborative approaches parallel those of EPA by involving industry, local governments, regional air pollution agencies, environmental, community, and public interest representatives. Participation ranges from informal advisory processes to formal negotiated rulemaking processes.

4.5 The Balancing Role for States Between Federal Directives and Business Regulation

Clean Air Act provisions since 1970 assign primary implementation functions to state governments. These include overall air quality planning, monitoring pollutant levels, regulating source mobile and stationary emission sources, and financing programs to meet federally determined standards. State program administrators coordinate their activities with EPA, multi-state coalitions, industries, substate districts, and local governments. They initiate and participate in technical advisory committees, partnership programs, and negotiated policy development. Managers also interact directly with industries, businesses emitting toxic pollutants, and with motor vehicle owners. Title V administration and finance create additional challenges. Independent of Constitutional concerns with this allocation of federal regulatory duties, states must balance the demands of federal air quality standards and direct relationships with regulated businesses.

[37] See Appendix, at pp. 122–126.
[38] See Appendix, at pp. 130–132.

5 Industry Responses to the Clean Air Act

Beyond restructuring governmental roles, the 1970 Clean Air Act created unyielding demands for industry compliance. National law authorized states to regulate both stationary and mobile sources pursuant to their air quality implementation programs. Industrial sites could be required to reduce emissions without regard for costs or current technological capacity. The 1990 amendments brought many smaller-scale businesses into federal jurisdiction due to their emissions of toxic pollutants.

Corporations responded to Clean Air Act demands by allocating necessary investments to equipment, research, and product development. Further, they escalated environmental concerns to top-level management. Over four decades, annual and specialized reports have evolved from reactive allegations of public sector encroachment toward proactive commitments to environmental sustainability.

The case studies in this chapter represent a cross-section of industrial sectors that have been critically impacted by air quality regulations: motor vehicle manufacturing (Ford Motor Company); chemical companies (DuPont, Rohm and Haas/Dow); oil refining operations (Exxon Mobil); semiconductors (Texas Instruments); consumer products (3M, Procter & Gamble); and aerospace (the Boeing Company). The primary data for the included case examples are annual and public affairs reports. These sources provide insights on corporate policies, capital and operating commitments, and changes in management structure that can be attributed to Clean Air Act compliance demands. Annual reports cite awards, cooperative public–private programs, and industry-wide initiatives such as the American Chemistry Association's Responsible Care program. Because their primary intent is to convey positive outlooks for stockholders, potential investors, and public relations, these reports have limited probative value. Nonetheless, each company has integrated environmental policy into their missions and high-level management structure over time.

The corporate documents referenced in this chapter reflect a trend away from wary initial reactions toward more proactive commitments to environmental sustainability goals. Through the 1970s, companies offered conflicting praise for the social objectives of environmental laws while pleading for policy makers to temper their regulations. Automotive, chemical, and other manufacturers focused advocacy on two fronts. First, they alleged that health-based Congressional mandates imposed disproportionate economic impact relative to the benefits provided to the public at large. Industries contended as well that they would be unable to develop or apply necessary emission-reduction technologies within imposed deadlines. Individual companies and their respective trade associations pressed for regulatory "reasonableness." They also reported significant capital expenditures and higher operating costs to improve, install, and develop control technologies to comply with the NAAQS.

An Interactive History of the Clean Air Act. DOI: 10.1016/B978-0-12-416035-4.00005-9

Table 5.1 Corporate Values—18 DJSI Sector Leaders

Values	DJSI Leading Companies
Sustainability	12
Responsibility	12
Integrity	11
Dialogue with stakeholders	8
Diversity	8
Innovation	10

Source: Ricart, Rodriguez, and Sanchez, "Sustainability in the Boardroom," at p. 6.

Industry criticisms waned after the 1977 amendments deferred compliance schedules with added deference to technological and economic concerns. The emphasis on limited government through the 1980s offered greater latitude; however, increased competition in the automobile industry brought greater pre-competitive collaboration to develop emission-control technologies and other efficiencies. Corporate participation through trade associations in EPA policy processes increased through the 1990s.

While some corporate goals can be linked directly to air pollution laws, other reporting merges environmental, social, and economic objectives into broader "citizenship" or "sustainability" reports. Another attribution factor concerns multinational corporations that aggregated their U.S. commitments within worldwide totals. In those situations, the cases refer either to overall expenditures for pollution control or to corporate policy and management shifts that encompass air quality management elements.

Potential cause–effect relationships between the Clean Air Act and industry responses became more abstract as companies adopted broader commitments to environmental sustainability (Table 5.1). Generalized commitments to minimize air, water, and toxic chemical pollution may not incorporate references to specific environmental laws. It would overstate the case to attribute these shifts to the impetus of the Clean Air Act. However, it is a reasonable inference to relate global climate change statements to present and emerging public policies for air quality management.

Environmental quality and sustainable development objectives are now well established among the companies profiled in this section. Corporations integrate pollution reduction, global warming concerns, and environmental technology within their operational goals. Announcements that environmental matters are board-level or top executive concerns are also prevalent. Table 5.2 highlights management, policy, and operational changes among the companies profiled in this section.

Beginning in the early 1970s, structural shifts in corporate governance prioritized environmental management by providing direct lines to top executives and boards of directors. DuPont created an Environmental Quality Committee headed by a senior vice president in 1971. This committee's vice chairman also served as corporate director of environmental affairs.[1] Rohm and Haas assigned comparable tasks to its vice president of engineering in that year. Corporate policy established environmental

[1] *See* E.I. DuPont de Nemours & Company, *DuPont Annual Report 1971*, at p. 19 (1972).

Table 5.2 Timeline of Corporate Administrative Changes Affecting Air Quality

1971	DuPont establishes an Environmental Quality Committee headed by a senior vice president.
	Rohm and Haas establishes environmental control as an engineering concept on par with structural design, safety, and technical process requirements.
1975	3M establishes innovative "Pollution Prevention Pays" (3Ps) program.
1977	Ford appoints executive vice president for Environmental, Safety, and Industry Affairs.
1981	Rohm and Haas adds internal review process for environmental controls and establishes an Environmental Advisory Council.
1984	DuPont appoints a vice president for Safety, Health, and Environmental Affairs.
1990	Texas Instruments declares its responsibility to protect the environment, primarily through aggressive waste minimization practices.
2001	Exxon Mobil structures environmental management within its Operations Integrity Management System (OIMS).
2004	Ford CEO created a vice president–level task force to establish a governance process for sustainable mobility.
2005	3M establishes corporate goal to reduce volatile air emissions 25% by 2010.
2007	Boeing rescales its corporate priorities by establishing an Environment, Health, and Safety division and policy council led by its president and CEO.
	Ford establishes new position of senior vice president of Sustainability, Environment, and Safety Engineering.
	P & G announces five-part sustainability strategy to be implemented by 2012.
2008	Ford establishes board-level Sustainability Committee.
2009	Exxon forms corporate-level sustainability working group.
2010	P & G establishes Global Sustainability Board and "sustainability vision."
	Texas Instruments releases multiyear sustainability goals.

control as an engineering concept on a par with structural design, safety, and technical process requirements. Exxon Mobil's pollution control programs are integrated within its company-wide Operations Integrity Management System (OIMS). In 2007, Boeing's president and CEO announced the creation of an Environment, Health, and Safety division that reports directly to him and the board's corporate policy council.

Cost estimates for Clean Act compliance have far exceeded EPA estimates. Members of the Alliance of Automotive Manufacturers (2005) reported that application to issuance costs for Title V permits averaged $170,000. This compared to the agency's estimate of $55,000. The Alliance estimated that ongoing compliance costs for assembly plants ranged from $150,000 to $300,000 annually.[2] Pharmaceutical company Eli Lilly estimated its "nonhidden" compliance costs for its three largest facilities at $7 million from 1994 to 2005.[3] In that same year, the Congressional Research Service reported that industries had not met the challenge of increasingly

[2] Title V Task Force, *Final Report to the Clean Air Act Advisory Committee: Title V Implementation Experience*, at p. 27 (April 2006). *See also* Michael Greenstone, "The Impacts Of Environmental Regulations On Industrial Activity: Evidence from the 1970 and 1977 Clean Air Act Amendments and the Census of Manufactures," *Journal of Political Economy*, 110 (2002): 1175–219.

[3] Ibid., at p. 28.

stringent regulations on emissions "... even when the technology to meet the standard had not necessarily been identified at the time the standards were initially promulgated."[4]

Despite continual criticism of regulatory burdens, some environmentally geared policies have provided opportunities for corporate growth. DuPont and Rohm and Haas established divisions to market emission-control technologies. DuPont also gained positive attention from its initiatives to control fluorocarbons in their products. 3M estimated that its Pollution Prevention Pays (3Ps) program saved the company nearly $1.2 billion through 2005. 3M and Procter & Gamble were early to develop and market consumer products to capture environmental concerns.

5.1 Ford Motor Company

With passage of the Clean Air Act, Ford Motor Company faced immediate challenges to develop and implement emission-control technologies within regulatory time frames. Its corporate reports through the 1970s and 1980s reference the demands for environmental compliance continually. Facing competition from foreign manufacturers, Ford entered pre-competitive research and development partnerships with Chrysler and General Motors by mid-decade. Through the 1990s, its corporate reports moderated in tone by highlighting environmental goals and commenting less on regulatory demands. Since 2002, Ford annual reports have emphasized environmental sustainability and related corporate responsibility matters.

Ford's 1970 annual report acknowledged Clean Air Act requirements to nearly eliminate hydrocarbon and CO emissions by the 1975 model year. However, the company told stockholders that government and industry had not accurately considered the impacts of rapidly expanding automobile use on air pollution in 1971:

> *We responded in what we thought was an appropriate manner at the time, but our efforts have not been adequate to forestall laws and regulations that sometimes go far beyond what is necessary, feasible, and economically reasonable.*[5]

The 1971 report urged the Congress to consider the inherent waste in a "crisis" approach. Toward this end, it recommended extending the time frame to comply with Clean Air Act emission standards beyond the 1975 model year.[6] The 1972 stockholder report called for "... realistic emission standards that allow adequate development and testing time and permit equipment and maintenance costs to be held in some reasonable balance with the benefits of air quality."[7] By 1977, Ford established a new position of executive vice president for Environmental, Safety, and Industry Affairs.[8]

[4] Stephen Cooney and Brent D. Yacobucci, *U.S. Automotive Industry: Policy Overview and Recent History*, CRS-95. Congressional Research Service, April 25, 2005.
[5] Ford Motor Company, *Ford 1971 Annual Report*, at pp. 2–3 (1972).
[6] Ford Motor Company, *Ford 1971 Annual Report*, at p. 3.
[7] Ford Motor Company, *Ford 1971 Annual Report*, at p. 3.
[8] *Ford Annual Report 1977*, at p. 23 (1978).

Ford's environmental tone moderated in the 1990s. The 1991 annual report set out a product challenge to accelerate development of alternative fuel vehicles that would contribute to a cleaner environment. In 1998, the CEO assured stockholders that business goals were not in conflict with social and environmental needs.[9] This report noted that continued improvement in environmental performance would also substantially reduce costs for the upcoming 5-year period.[10]

In 2000 and 2001, Ford continued developing a set of business principles with active employee participation. The 2001 annual report, entitled *Connecting With Society: Our Learning Journey*, affirmed a company-wide commitment to environmental protection.[11] It described a new coordinative structure among the Environment and Safety Engineering, Corporate Governance, and Public Affairs offices.[12]

Ford's *2002 Corporate Citizenship Report* acknowledged that the automobile industry and those who use its products are significant generators of GHG emissions. However, it referred to climate change as a societal challenge to be confronted by all stakeholders. Similar to other companies, Ford advocated "sharing the responsibilities and burdens of reducing greenhouse gas emissions with continuing sustained economic growth."[13] The 2002 citizenship report also cited emerging relationships with automotive, manufacturing, trade associations, and business organizations focused on sustainability.

In 2004, Ford's CEO created an executive task force to establish a governance process for sustainable mobility, climate change policy, and technologies to improve fuel economy.[14] The Board of Directors Environmental and Public Policy Committee was assigned primary strategic responsibility for these functions. Ford created a new position of senior vice president of Sustainability, Environment and Safety Engineering in 2007.[15]

The *2008 Annual Report* referenced a "sustainability strategy that outlines future technologies pathways for ... vehicle production in the near, mid and long term." The "Drive Green" program committed to producing engines with 20% greater fuel economy and a 15% reduction in carbon emissions.[16] The company's 2009–2010 *Sustainability Report* highlighted a commitment to "effective and appropriate" climate change policy. It also referenced Ford's commitment to the consensus recommendations of the U.S. Climate Action Partnership (USCAP) [discussed in Chapter 6] as a factor helping to transform the company's product line.[17]

[9] *Ford Motor Company 1998 Annual Report*, at p. 2 (1999).

[10] *Ford Motor Company 1998 Annual Report*, at p. 2.

[11] Ford Motor Company, *Connecting With Society: Our Learning Journey*, at pp. 12–13. Ford Motor Company, 2008.

[12] Ibid., at p. 62.

[13] Ford Motor Company, *2002 Corporate Citizenship Report*, at p. 36 (2003).

[14] Ford Motor Company, *2007/8 Blueprint for Sustainability* 9 (2008).

[15] *See* Ford Motor Company, *2007/8 Blueprint for Sustainability*, at p. 8 (2007).

[16] Ford Motor Company, *2008 Annual Report*, at p. 14.

[17] Ford Motor Company, *2009–2010 Sustainability Report*, at p. 3 (2010).

5.2 DuPont

The E.I. DuPont & Nemours has consistently adapted to air and water pollution laws while challenging their reasonableness. Its 1970 report to stockholders identified the increasingly significant impact of pollution control laws as a business expense.[18] It stated further that "little of this investment yields a direct dollars-and-cents return."[19] In that year, approximately 1,500 of DuPont's employees worldwide focused on environmental matters.

Within the United States, DuPont expended $168 million on operating and investment costs by 1970.[20] The company adapted its organizational structure the following year (1971) by establishing an Environmental Quality Committee headed by a senior vice president. The committee's vice chairman served as corporate director of environmental affairs.[21] Environmental quality representatives from each industrial department provided staff support and met monthly to coordinate policy implementation.[22] DuPont estimated that 1,700 full-time employees were devoted to pollution control activities during 1971.[23]

Staff commitments to environmental compliance increased to 2,000 in 1972.[24] In that year, DuPont committed $72 million to the development of pollution abatement facilities (air, water, and noise) within the United States. That constituted 8% of its domestic construction costs.[25] Costs for managing facilities and for environmental research and development in 1973 was $92 million with 2,450 full-time employees.[26] The company also reported that its internal compliance technologies were being marketed to other companies addressing pollution control.[27] By 1974, 2,900 employees were managing a total environmental budget of $120 million.[28]

DuPont's reports through the 1970s expressed grave concerns over the fiscal impacts of company compliance with environmental regulations. The 1976 annual report expressed apprehension about ongoing compliance costs in relation to potential environmental benefits. This reflected a broader concern among industries "to press for corrective amendments to existing U.S. legislation and for a more realistic and balanced approach to new legislation and regulations."[29]

In 1979, DuPont declared a renewed optimism based on the EPA's response to industry advocacy for a "bubble" approach to air pollution control. This allowed multiple sources within a fixed area to be treated as one emission source with cost reductions as high as 40%.[30] DuPont's 1980 report expressed intent to increase

[18] *E.I. DuPont & Nemours Company Annual Report 1970*, at p. 4 (1971).
[19] Ibid.
[20] Ibid., at p. 4.
[21] *DuPont Annual Report 1971*, at p. 19 (1972).
[22] Ibid., at pp. 14, 19.
[23] Ibid., at p. 19.
[24] *DuPont Annual Report 1972*, at p. 14 (1973).
[25] Ibid., at p. 14.
[26] *DuPont Annual Report 1973*, at p. 17 (1974).
[27] Ibid., at p. 14.
[28] *DuPont Annual Report 1974*, at p. 20 (1975).
[29] *DuPont Annual Report 1976*, at p. 18 (1977). *See also DuPont Annual Report 1975*, at p. 12.
[30] *DuPont Annual Report 1979*, at p. 24 (1980).

efficiency in its environmental control programs and "to participate effectively in the formation of public opinion and policy leading to environmental regulations."[31] The 1982 Annual Report referenced continuing legislative efforts on the Clean Air Act.[32] The company appointed a vice president for Safety, Health, and Environmental Affairs in 1984 to manage these programs more effectively.[33]

DuPont's 1988 report announced a corporate decision to phase out production of halogenated fluorocarbons by 2000.[34] The stockholder report for 1989 acknowledged growing public concern and corporate commitment toward "environmental stewardship."[35] It references the company's commitment along with other chemical manufacturers to the Responsible Care® program.[36]

DuPont capitalized on its industry status by establishing a new business unit for Safety and Environmental Resources. This division markets environmental services to customers and other potential users.[37] The EPA awarded the company its Stratospheric Ozone Protection Award in 1990 for leadership role in phasing out CFCs.[38] The 1991 report asserted to stockholders that "it would be irresponsible to disregard the considered positions of international and world bodies through a unilateral decision to cease production."[39] In 1996, the CEO referred to an emerging phase of environmental performance as a "business requirement for competitiveness."[40] The 1999 annual report described "sustainable growth" as an objective that would increase shareholder value while reducing the company's environmental footprint.[41]

Since 2000, DuPont has emphasized its commitment to voluntarily reducing its environmental footprint. The 2005 annual report, entitled *Sustainable Growth Through Science*,[42] restated this objective as a primary corporate mission.[43] In 2006, the Chairman of the Board described the company's emerging sustainability focus as a market-driven business fundamental:

> *We see ourselves entering a new phase of sustainability. The first was a focus on internal safety and meeting environmental regulations back in the 1970s. In the late 1980s and 1990s came voluntary foot-print reductions, going beyond regulatory requirements and pursuing a goal of zero safety and environmental incidents. Now we are in a third phase of sustainable growth, characterized by a holistic approach, fully integrated into our business models.*[44]

[31] *DuPont 1980 Annual Report*, at p. 20 (1981).
[32] *DuPont 1982 Annual Report*, at p. 3 (1983).
[33] *DuPont 1984 Annual Report*, at p. 20 (1985).
[34] *DuPont: A Global Entrepreneur Annual Report 1988*, at p. 3 (1989).
[35] *DuPont Annual Report 1989*, at p. 6 (1990).
[36] *DuPont Annual Report 1989*, at pp. 6–7.
[37] *DuPont Annual Report 1989*, at p. 20.
[38] *DuPont Annual Report 1990*, at p. 4 (1991).
[39] *DuPont Annual Report 1991*, at p. 25 (1992).
[40] *DuPont 1996 Annual Report*, at p. 6 (1997).
[41] *DuPont 1999 Annual Report*, at p. 2 (2000).
[42] Dupont, *Sustainable Growth Through Science—2005 Annual Review* (2006).
[43] DuPont, *2015 Sustainability Goals*, at p. 2 (2006).
[44] *DuPont 2006 Annual Review*, at p. 3 (2007).

DuPont described its sustainable growth mission in 2007 as creating shareholder and societal value while reducing its environmental footprint in its operating sectors.[45] In a preface to the 2010 sustainability report, its CEO described the following benefits from the company's sustainability commitment:

Recently, we engaged customers around the world to determine their value for sustainable products, understand their market drivers, and assess the longevity of environmental trends and the potential for green job creation. More than 89 percent of those surveyed reported that customer demand is a key driver for developing products with an enhanced environmental profile, that overwhelmingly there is value for environmental benefits in products now, and that value will only continue to increase in the coming years.[46]

Compared to four decades earlier, DuPont's 2010 annual report offers a collaborative tone toward "building alliances with people, companies, governments and organizations around the world in an effort to improve the lives of people everywhere."[47]

5.3 Rohm and Haas

Rohm and Haas, a wholly owned subsidiary of the Dow Chemical Company since 2009, produces specialty materials ranging from paint coatings and sunscreens to semiconductor chips. In 1970, the company began its transition from reliance on local management and assistance to assigning corporate responsibility and long-term environmental compliance planning to its vice president of engineering.[48]

By 1971, Rohm and Haas established environmental control as an engineering concept on par with structural design, safety, and technical process requirements.[49] Even at that early stage, company ecologists and other scientists maintained direct contact with federal, state, and local government officials.[50] The 1971 report also expressed support for legislation that provides constructive standards for pollution control.

Among significant expenditures for 1971, Rohm and Haas converted the powerhouse at two industrial sites from coal to oil to reduce SO_2 emissions.[51] In 1972, it reported increased cooperation with government authorities in developing environmental guidelines and controls.[52] The company's overall environmental expenditures

[45] *Dupont 2007 Annual Review*, at p. 7 (2008).
[46] Ellen Kullman, "A Message from our CEO," in DuPont, *2010 Sustainability Progress Report*, at p. 3 (2010).
[47] DuPont, *2010 Annual Review*, at p. 3 (2011).
[48] *Rohm and Haas Annual Report 1970*, at p. 3 (1971).
[49] *Rohm and Haas Annual Report 1971 Report*, at p. 10 (1972).
[50] *Rohm and Haas Annual Report 1971 Report*, at p. 10.
[51] Ibid.
[52] *Rohm and Haas Company Annual Report 1972*, at p. 13 (1973).

in 1973 increased to $18.2 million.[53] It installed scrubbers at two industrial sites in 1974 to improve control over sulfur oxide and particulates.[54]

The Rohm and Haas 1976 annual report expressed willingness to comply with, or exceed, environmental regulation. The company invested $2.4 million in that year to install scrubbers to meet EPA standards for emitting SO_2 gas.[55] It took further control measures by adding fiberglass filters to remove fine mist droplets with ammonia or acid from waste gases.[56]

In 1979, Rohm and Haas reported that its environmental control programs were current with regulations, which allowed the company to prepare for future trends.[57] Two years later, it added an internal review process with teams of three to six technical staff assessing each facility's compliance with corporate and government requirements.[58] Further, it created an Environmental Advisory Council comprised of two nonmanagement directors and two outside consultants reporting directly to the Board of Directors.[59]

Rohm and Haas continued its leading-edge environmental policies through the 1990s. In 1992, it committed to a goal of 75% reduction in air emissions by 1996.[60] While achieving only a 50% reduction during this period, it had expanded its plant production by more than 60%.[61] The company added sustainable development to its guiding principles in 1997.[62] The EPA highlighted company efforts in 1998 by offering its Green Chemistry Award.[63] In 2000, the agency cited three of Rohm and Haas plants as among the first whose performance and management exceeded regulatory compliance "to the benefit of people and the local environment."[64]

By 2006, Rohm and Haas had integrated sustainability into its overall mission statement. The *2006 EHS and Sustainability Report* asserts that: "[s]ustainable development and environmental excellence are the cornerstones of our product development and manufacturing processes."[65] It set long-term goals to attain "zero discharges" and "robust compliance with external regulations and voluntary commitments."[66] In January 2007, the company committed to "meet or exceed all applicable laws, regulations, and Rohm and Haas standards ..." and "strive to prevent or reduce pollution from emissions, discharges, and wastes."

Rohm and Haas became a subsidiary of Dow Chemical in April 2009. The parent company's annual report for that year stated its commitment to connect

[53] *Rohm and Haas Company Annual Report 1973*, at p. 18 (1974).
[54] *Rohm and Haas Company 1974 Annual Report*, at p. 19 (1975).
[55] *Rohm and Haas 1976 Annual Report*, at p. 18 (1977).
[56] Ibid.
[57] *Rohm and Haas 1979 Annual Report*, at p. 19 (1980).
[58] *Rohm and Haas Annual Report 1981*, at p. 26 (1982).
[59] *Rohm and Haas 1985 Annual Report*, at p. 31 (1986).
[60] *Rohm and Haas Annual Report 1992*, at p. 22 (1993).
[61] *Rohm and Haas Annual Report 1995*, at p. 19 (1996).
[62] *Rohm and Haas Annual Report 1997*, at p. 19 (1998).
[63] *Rohm and Haas Annual Report 1998*, at p. 21 (1999).
[64] *Rohm and Haas Annual Report 2000*, at p. 23 (2001).
[65] *Rohm and Haas Annual Report 2003*, at p. 12 (2004).
[66] *Rohm and Haas 2006 EHS and Sustainability Report*, at p. 1 (2007).

"... chemistry and innovation with the principles of sustainability ...."[67] Dow has described its sustainability goals as "integral to its corporate vision, mission, and values ... which continue to drive change that is good for the environment, good for people and good for business."[68] In 2010, Dow initiated a dual-pronged sustainability strategy that incorporates both business and citizenship elements.[69]

5.4 Exxon Mobil

The Exxon Corporation merged with the Mobil Oil Company in 1999. Prior to its rebranding as Exxon in 1973, Standard Oil of New Jersey's 1970 annual report acknowledged a need to "work with others toward a consensus on physical environmental qualities which are desirable and attainable." However, this acknowledgment was limited to conducting an "extensive study both to the definition of environmental conservation problems encountered in business and to the development of reasonable solutions."[70]

Exxon's 1975 annual report was the first to directly reference environmental protection measures. It cited over $2.8 billion in environmental conservation costs from the previous decade. Its $178 million in that year's costs included $35 million for wet gas scrubbers and $70 million for facilities to reduce pollutants—primarily sulfur—from its refined products.[71] In 1976, overall operating costs for environmental protection were $685 million. This annual report also referenced economic benefits through the licensing of its innovative technologies for removing criteria pollutants. These included its wet scrubber processes for removing particulates and most sulfur oxides, and for reducing nitrogen oxide emissions from large boilers.[72]

Exxon's stated commitments to air pollution control continued through the 1990s. In 1993, the company spent $33 million to install fuel vapor recovery systems at 800 service stations.[73] It introduced reformulated gasoline and upgraded its refineries to produce lower-sulfur diesel fuels in 1995. Exxon also reported its participation in joint petroleum and auto industry research programs to pursue improvements in engine and fuel technologies.[74] The 1995 report noted further that these "industry-funded programs develop a sound scientific base to assist regulators in defining cost-effective routes to environmental protection."[75]

Before merging with Mobil Oil Company, Exxon's 1998 report asked for "common sense" and "voluntary market-driven" strategies on the issue of global climate change.[76]

[67] Dow Chemical Company, *2009 Annual Report*, at p. 11 (2010).
[68] Dow Chemical Company, *2010 Annual Report*, at p. 11 (2011).
[69] Dow, *2010 Global Reporting Initiative Report/Annual Sustainability Report*, at p. 6 (2010).
[70] *Standard Oil Company (New Jersey) 1970 Annual Report*, 14 (1971).
[71] *Exxon Corporation 1975 Annual Report*, at p. 19 (1976).
[72] *Exxon Corporation 1976 Annual Report*, at p. 18 (1977).
[73] *Exxon Corporation 1993 Annual Report*, at p. 6 (1994).
[74] *Exxon Corporation 1994 Annual Report*, at pp. 4–5 (1995).
[75] *Exxon Corporation 1995 Annual Report*, at p. 6.
[76] *Exxon Annual Report 1998*, at p. 5 (1999).

It urged further that: "[i]n the longer term, decisions must be guided by sound science ..." and that "economic health contributes to both social and environmental well-being."[77]

Exxon Mobil's 2000 report continued to recommend that public policies rely "... primarily on scientific and economic analysis to develop feasible, economically justified and effective recommendations."[78] The 2006 Citizenship Report expressed a strategic commitment to reducing GHG emissions through research and development of lower-emission technologies. However, that report remained noncommittal on whether there is an underlying scientific rationale:

> *Climate remains an extraordinarily complex area of scientific study. Nevertheless, the risk to society and ecosystems from rising greenhouse gas emissions could prove to be significant. So, despite the areas of uncertainty that exist, it is prudent to develop and implement strategies to address this risk.*[79]

An investor accountability consortium criticized the 2006 report's position on climate change science.[80] Exxon Mobil responded moderately by acknowledging increased political attention to climate change impacts in its *2007 Corporate Citizen Report*. It also called for "increased awareness of how energy shapes our world as well as discussions on policies that seek to reduce greenhouse gases."[81] The company's 2008 report reflected a comparable commitment along with reservations over potential policy and control strategies:

> *A number of organizations have attempted to quantify the potential implications of climate-related policies for oil and gas industry shareholders. However, these efforts are based on regulatory assumptions that are only speculative given the current status of negotiations on climate-related policies.*[82]

The 2009 annual report cites Exxon Mobil's $1.9 million in cogeneration technology and other activities that reduce GHG emissions in comparison to conventional power generation.[83]

Exxon Mobil's environmental policy is structured within its 11-point OIMS (Figure 5.1). This looping process integrates air pollution control with other management decisions at the corporate as well as operational level. It also seeks a balance between environmental and economic needs. Beyond compliance with applicable laws and regulations, Element Four commits to "responsible standards where laws and regulations do not exist."[84] This includes working with government and industry

[77] *Exxon Annual Report 1998*, at p. 5 (1999).

[78] *Exxon Mobil Summary Annual Report 2000*, at p. 6 (2001).

[79] Exxon Mobil, *2006 Citizenship Report*, at p. 3 (2006).

[80] Andrew Logan and David Grossman, Exxon Mobil's Corporate Governance on Climate Change, at pp. 2–7. CERES, 2006.

[81] Exxon Mobil, *2007 Corporate Citizenship Report*, at p. 3 (2007).

[82] Exxon Mobil, *2008 Corporate Citizenship Report*, at p. 30 (2009).

[83] Exxon Mobil, *2009 Summary Annual Report*, at p. 10 (2010).

[84] Exxon Mobil, *Managing for Environmental Excellence*, at p. 8 (November 2004).

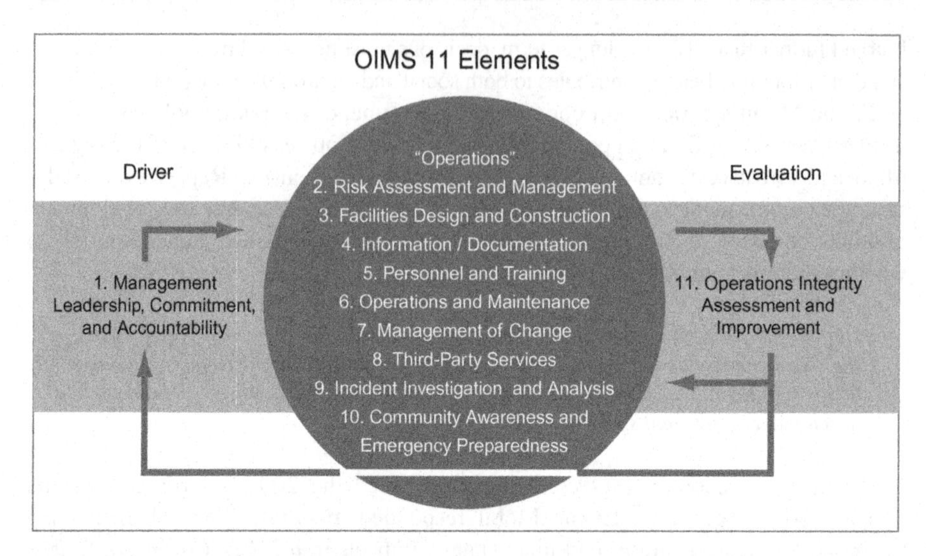

Figure 5.1 Exxon Mobil, Operations Integrity Management System.
Source: Exxon Mobil, OIMS Report (2009).

groups "to foster timely development of effective environmental laws and regulations based on sound science ...." Other environmental policy elements are "controlling emissions and wastes to below harmful level" and enhancing capacity to make operations and its products compatible with the environment.[85] In a 2009 report, Exxon Mobil's CEO refers to the OIMS framework as a key element in improved emission controls.[86] The company established a corporate-level sustainability working group in that year.[87]

5.5 Texas Instruments

In 1990, Texas Instruments declared its responsibility for doing its share to protect the environment. This commitment focused primarily on aggressive waste minimization practices.[88] The company's 1991 annual report cited its role as a founding member of the Industry Cooperative for Ozone Layer Protection and as a volunteer in the EPA's initiative to reduce emissions from 17 high-priority chemicals.[89] In 1994, company efforts to eliminate ozone-producing substances received the Stratosphere Ozone Protection Award from the EPA.[90]

[85] Ibid., at p. 9.
[86] "Chairman's Message," in Exxon Mobil, *Operations Integrity Management System*, at p. 3 (2009).
[87] Exxon Mobil, *2009 Corporate Citizenship Report*, at p. 2 (2009).
[88] *Texas Instruments Incorporated 1990 Annual Report*, at p. 11 (1991).
[89] *Texas Instruments 1991 Annual Report*, at p. 5 (1992).
[90] *Texas Instruments 1994 Annual Report*, at p. 22 (1995).

Texas Instruments' Worldwide Environmental, Safety and Health Director affirmed in 2004 that: "TI has long set aggressive goals aimed at reducing any environmental impact from our products and operations ...."[91] This statement referred to the company's early decision to move toward lead-free products.[92] The 2005 environmental report cites its commitment to a multidisciplinary precursor control program for NO_x emissions.[93]

Similar to other companies, Texas Instruments has brought sustainability into its corporate mission and operations:

> *At TI, sustainability is more than just a buzzword. It is a way of life. TI has taken a multi-faceted approach to environmental sustainability in areas where it makes good sense for TI's bottom line and the environment and has long pursued the goal of "zero wasted resources."*[94]

These statements also acknowledge that actions are necessary for its license to operate at several key manufacturing sites.[95]

Citizenship reports from 2008 to 2010 highlight corporate activities relevant to air quality. In 2008, TI received the EPA's Clean Air Excellence Award and State of Texas recognition for technological innovation in volatile organic compound (VOC) emissions from semiconductor manufacturing processes.[96] The company announced in 2009 that perfluorinated chemical (PFC) emissions were 10% less than the 1995 baseline, and that it had reduced nitrogen oxide emissions by closing down old equipment and using technologically advanced abatement systems.[97] Policies on GHG reporting or regulatory demands focused on potential costs that could put the company at a competitive disadvantage.[98]

5.6 3M Corporation

The 3M Corporation has integrated environmental compliance into its corporate policies and practices since 1971 when its annual report stated a commitment to "meet or exceed pollution control standards at all manufacturing locations."[99] The company's largest capital expenditure in that year was the construction of a multimillion dollar advanced incineration system at its Cottage Grove, Minnesota

[91] Texas Instruments, *Environmental, Safety, and Health—2004 Review*, at p. 6 (2005).
[92] Texas Instruments, *Environmental, Safety, and Health—2004 Review*, at p. 6.
[93] Texas Instruments, *Building a Better Future—Environmental, Safety, and Health 2005 Review*, at p. 6 (2006).
[94] Texas Instruments, *Building a Better Future*, footnote 222, at p. 6.
[95] Texas Instruments, *Building a Better Future*, footnote 222, at p. 6.
[96] Texas Instruments, *2008 Corp Citizenship Report Summary*, at p. 10.
[97] Texas Instruments, *2009 Corp Citizenship Report Summary*, at p. 10.
[98] Texas Instruments, "Public Policy Priorities." http://www.ti.com/corp/docs/company/citizen/government/key.shtml (accessed August 2011).
[99] *3M Company 1971 Annual Report*, at p. 24.

plant complex.[100] The 1973 annual report expressed further emphasis on product research and development to reduce pollution in manufacturing processes and by customers.[101]

The 3M established its 3Ps program in 1975 with the premise that "a prevention approach is more environmentally effective, technically sound, and is more economical than conventional pollution control equipment."[102] The 3P strategy seeks to improve the environmental efficacy of its industrial processes and to eliminate reliance upon raw materials that "tend to be pollutants or have questionable toxicity."[103] The company affirmed the wide-ranging benefits of its 3P program in a 1976 conference cosponsored by the EPA including "a cleaner environment, lower investment in pollution control equipment, savings in energy and raw material costs, and perhaps, sales that might otherwise have been lost."[104] In 1983, 3M reported total savings from the 3P program at $190 million.[105]

Through 2005, estimated cost savings from the 3Ps program were nearly $1.2 billion. This was achieved primarily through improved products and manufacturing processes rather than removing pollutants after-the-fact.[106] The company also offered individual awards to employee projects that eliminate or reduce pollutants, improve manufacturing efficiency, or save costs by avoiding or deferring pollution (Figure 5.2).[107] Its current organizational structure reflects significant corporate attention to sustainability. An executive-level Corporate Environmental, Health and Safety Committee and its Business Conduct Committee report to the chairman, president, and CEO. Staff from environmental and other departments provide technical support.

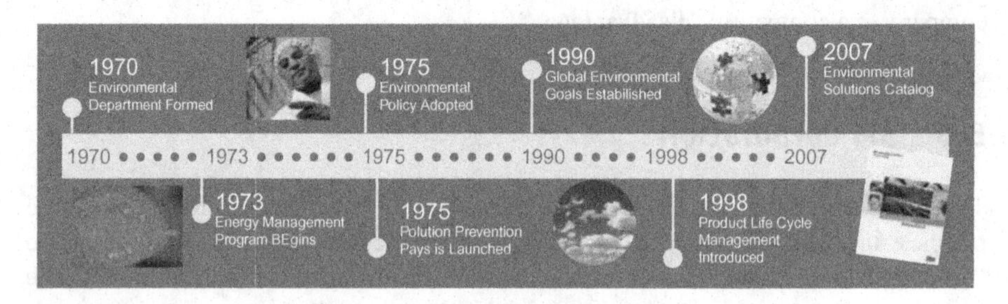

Figure 5.2 3M timeline for sustainability.
Source: 3M, Environmental Solutions Catalog (2009).

[100] *3M Company 1971 Annual Report*, at p. 24.
[101] The 1973 report also references installation of additional thermal oxidizers for its pressure-sensitive tape plant in Bedford Park, Illinois. *3M Company 1973 Annual Report*, at p. 31 (1974).
[102] *See* http://solutions.3m.com/wps/portal/3M/en_US/global/sustainability/management/pollution-prevention-pays (accessed October 2011).
[103] *3M Company1977 Annual Report*, at p. 5 (1978).
[104] *3M Company 1976 Annual Report*, at p. 3 (1977).
[105] *3M Company 1983 Annual Report*, at p. 8 (1984).
[106] *3M Company 1983 Annual Report*, at p. 8 (1984).
[107] *3M Company 1983 Annual Report*, at p. 8 (1983).

The 3M's sustainability strategy links economic success to environmental stewardship and social responsibility.[108] It introduced an *Environmental Solutions Catalog* in 2007 as part of its broader sustainability program to showcase products for customers to improve energy efficiency and reduce GHG emissions.[109] A 2008 report entitled *Sustaining Our Future* links long-term corporate success to adoption and implementation of the following sustainable development principles: "stewardship to the environment, contributions to society, and to the creation of economic value and worth." Another significant element in an evolving sustainability strategy is to engage stakeholders with diverse viewpoints. 3M considers these processes as a means to learn from what succeeded or failed in the approaches of other companies.[110]

5.7 The Procter & Gamble Company

The Procter & Gamble Company (P & G) 1970 annual report stated at the outset that it was "wholly committed to eliminate Company-caused sources of pollution."[111] Its initial emission-control efforts included multimillion dollar investments to design and construct stock beaters, cyclone separators, and electrostatic precipitators.[112] The company reported technological advances in 1972 and made a commitment to help find viable solutions by sharing its methods and research findings with private and public agencies.[113] From 1993 through 1995, P & G reported 75% reductions in air and water emissions at its manufacturing plants, according to EPA-tracked results.[114]

In 1999, P & G's first annual *Sustainability Report* announced the appointment of a Director—Corporate Sustainable Development. It also directed the Corporate Environmental Quality Group to integrate economic progress and social development issues into its charge. P & G also restructured its overall functions into Global Business Units that were directed to integrate sustainable development into their business plans, work processes, and culture.[115] These changes reflected company acknowledgment of the emerging significance of sustainable development as both public policy issue and business opportunity.[116] The 1999 *Sustainability Report* also noted P & G's participation in industry associations for chemical producers and manufacturers, as well as the Global Environmental Management Initiative.[117]

[108] 3M, "About 3M Sustainability." http://solutions.3m.com/wps/portal/3M/en_US/3M-Sustainability/Global/VisionHistory/About/ (accessed October 2011).
[109] 3M, *2009 Environmental Solutions Catalog*, at p. 5 (2009).
[110] 3M, *2008 Sustainability Progress*, at p. 4 (2008).
[111] *The Procter & Gamble Annual Report for the Year Ended June 30, 1970*, at p. 9 (1970).
[112] Ibid., at p. 11.
[113] *Procter & Gamble Annual Report for the Year Ending June 30, 1972*, at p. 17 (1972).
[114] *Procter & Gamble, Global Opportunities Global Growth, 1995 Annual Report*, at p. 17 (1995).
[115] Procter & Gamble, *1999 Sustainability Report*, at p. 3 (1999).
[116] Ibid., at p. 7.
[117] Ibid., at p. 9.

The P & G 2000 sustainability report noted the absence of clear corporate examples for success using sustainable development as a basic concept.[118] The following year's report acknowledged growing scientific evidence linking GHGs to global climate change. This discussion urged that regulatory actions "provide maximum flexibility ... to minimize negative economic impacts on countries, on individual businesses, and ultimately on the general public."[119]

P & G's current policy and organizational structure places sustainable development within its international core values and guidelines. The company established a Global Sustainability Board in 2010 led by the North American Group president. Other members include officers from the Global Technology, Product Supply, Brand Building and External Relations Officers, and the North America Group president.[120]

5.8 Boeing

The Boeing Company's 1996 annual report references 5 years of initiative in developing and applying materials and processes that are environmentally sensitive.[121] It announced 58% reductions in chemical emission reductions of 58% within this time frame.[122] In 1998, Boeing joined the Business Environmental Leadership Council of the Pew Center for Global Climate Change.[123] The company reported as well that it had received a special recognition award from the EPA's Climate Wise program to reduce GHG emissions.[124]

Boeing rescaled its corporate priorities in 2007 by establishing an Environment, Health, and Safety division and Policy Council led by its president and CEO. The Council's mission is to set performance targets and to measure the effectiveness of these corporate strategies.[125] The report to stockholders emphasized that environmental strategy and performance would be "monitored at the highest levels of company leadership."[126] This includes company-wide environmental management systems and risk management. The tracking system extends throughout its design and operating processes, including compliance by Boeing's business partners.[127]

In Boeing's *2008 Environment Report*, the newly appointed vice president for Environment, Health, and Safety affirmed a clear mandate "to further embed

[118] Procter & Gamble, *2000 Sustainability Report*, at p. 4 (2001).
[119] Procter & Gamble, *2000 Sustainability Report*, at p. 19.
[120] "Leadership Statement," in Procter & Gamble, *2010 Sustainability Report*, at p. 5 (2010).
[121] *The Boeing Company, 1995 Annual Report*, at p. 23 (1996).
[122] *The Boeing Company, 1995 Annual Report*, at p. 23.
[123] *The Boeing Company, 1998 Annual Report*, at p. 20 (1999).
[124] *The Boeing Company, 1998 Annual Report*, at p. 20.
[125] *The Boeing Company 2007 Annual Report*, at p. 18 (2008).
[126] *The Boeing Company, 1998 Annual Report*, at pp. 3–4.
[127] *See* http://www.environmentalleader.com/2007/05/07/boeings-armstrong-heads-new-enterprise-environmental-organization (accessed August 2011).

environmental performance into Boeing's thinking, culture, and action."[128] She stated that: "Boeing has set aggressive and transparent enterprise-wide performance targets to drive environmental thought and action throughout its operations."[129] This report also announced a new policy to disclose publicly the company's carbon footprint and hazardous waste statistics and ongoing commitment to regulatory compliance.[130] Boeing also joined the EPA Climate Leaders program in 2008. This commits the company to conducting an inventory of GHG emission, establishing target reduction levels, and providing annual progress reports to EPA.[131] The company's internal goals toward 2012 are to reduce energy use, intensity of GHG emissions, and hazardous waste by 25%.[132]

5.9 The Impact of Expanded Federal Regulations on Small-Scale Stationary Sources

The 1990 Clean Air Act amendments extended the scope of implementation by applying technology-based standards to sources unfamiliar with national regulatory controls. For example, approximately 2,600 commercial bakeries within the United States would be responsible for limiting ethanol and other toxic emissions.[133] Federal criteria could now direct fermentation processes in dough production, VOCs emissions from ovens, cooling boxes, and packaging. Recommended control technologies could include oxidizers applying catalytic, thermal, and regenerative oxidizers.[134] The EPA also initiated regulations for perchloroethylene emissions from approximately 25,000 dry cleaning facilities nationwide.[135] A 1995 compliance manual recommended operation and machine improvements that ranged from closing dryer doors immediately, periodic lint cleaning, and duct repair to major equipment upgrades.[136]

5.10 Sustainable Development as an Emerging Element in Corporate Culture

The case studies in this chapter illustrate an overall trend from resistance to Clean Air Act requirements to ongoing corporate sustainability programs. Environmental

[128] Mary Armstrong, "Message from Mary Armstrong, Vice President for Environment, Health, and Safety," *Boeing 2008 Environment Report*, at p. 6 (2008).
[129] Mary Armstrong, "Message," footnote 255, at p. 7.
[130] Mary Armstrong, "Message," footnote 255, at p. 7.
[131] *Boeing 2008 Environment Report*, at p. 5 (2008).
[132] *Boeing 2008 Environment Report*, at p. 24.
[133] Of the 600 large-scale facilities, 23 had installed emission control devices. EPA, *Alternative Control Technology Documents for Bakery Oven Emissions*, at pp. 1–2 (1992).
[134] Ibid., at pp. 1–4, 5.
[135] "EPA Regulates Dry Cleaners in First Air Toxics Rule Under New Clean Air Act," EPA Press Release, September 14, 1993.
[136] EPA Office of Compliance Sector Notebook Project, *Profile of the Dry Cleaning Industry*, at pp. 40–2. EPA, September 1995.

affairs have risen to top-level management and board of director committees. Individual companies also cooperate in industry-wide programs and partnerships with government programs. Chapter 6 addresses the role of trade associations in assisting with industry–government relations.

5.11 Corporate Advocacy in Climate Change Policies

Each corporation covered in this chapter has developed programmatic responses and communication strategies for climate change issues. Dow, DuPont, and Ford are part of the United States Climate Action Partnership (USCAP). This alliance of business and environmental groups advocate federal legislation to require significant reductions of GHG emissions (See Chapter 6). All eight companies provide annual updates to the Carbon Disclosure Project, a voluntary global climate change reporting system.[137]

The policy statements in Table 5.3 indicate preferences for market-based approaches. They also acknowledge that voluntary measures may not be sufficient. In such cases, there is a preference for some form of monetization. Among potential regulatory options, the two prevalent ones are through taxation or via a "cap and trade" strategy. The latter approach would extend the "bubble concept" to the national scale by allowing entities emitting GHGs above specified limits to purchase pollution rights from others who have exceeded applicable criteria.

Table 5.3 Company Positions on Global Climate Change Issues

3M	3M endorses a voluntary, market-based approach that involves all nations; ... [with] provisions for emissions trading and credit for early action.[a]
Boeing	As the global community develops approaches to reducing GHG emissions, Boeing acknowledges that voluntary measures alone may not be enough and supports development of mandatory yet flexible frameworks to address emission reductions.[b]
Dow Chemical Company [Rohm and Haas]	Dow will advocate for and participate in the monetization of carbon in fair marketplaces, a critical objective in establishing country market mechanisms for cost-effective carbon management. Each country should be allowed to establish their own systems with targets set fairly for each industry sector.[c]
DuPont	We believe the scientific understanding of climate change is sufficient to compel prompt, effective actions to limit emissions of GHGs. We believe that to be successful these actions will require concerted engagement by the world's governments, along with technological innovations by businesses, and individual actions by all citizens ....[d]

(Continued)

[137] *See* https://www.cdproject.net/en-US/Pages/HomePage.aspx (accessed October 2011).

Table 5.3 (Continued)

Exxon Mobil	When considering policy options, Exxon Mobil advocates an approach that: • Ensures a uniform and predictable cost of carbon across the economy; • Lets market prices drive the selection of solutions; • Maximizes transparency to companies and consumers; • Reduces administrative complexity; • Promotes global participation; and • Is easily adjusted to future developments in climate science and the economic impacts of climate policies. A well-designed carbon tax is better able to accommodate these key criteria than other alternatives, such as cap and trade .... Combined with further advances in energy efficiency and new technologies spurred by market innovation, such a carbon tax could play a significant role in addressing the challenge of rising GHG emissions.[e]
Ford	We believe we need a comprehensive, market-based approach to reducing GHG emissions if the United States is going to reduce emissions at the lowest cost per ton. An economy-wide program would provide flexibility to regulated entities while allowing market mechanisms to determine where GHG reductions can be achieved at the lowest cost.[f]
Procter & Gamble	We believe that industry, government, and consumers all have roles to play in addressing climate change and that prudent and cost-effective actions to reduce GHG emissions to the atmosphere are necessary.[g]
Texas Instruments	We closely track global energy and environmental concerns, and we are committed to being part of the solution. In addition, we work through associations to provide context and perspective on the potential impacts of legislative and regulatory proposals.[h]

[a]3M, "Greenhouse Gas Management Policy." http://mws9.3m.com/mws/mediawebserver.dyn?6666660Zjcf6lVs6EVs66SfofCOrrrrQ- (accessed October 2011).
[b]Boeing, "Boeing's Environmental and Climate Change Policies." http://mdc.com/aboutus/environment/policies.html (accessed October 2011).
[c]Dow Chemical Company, "Addressing Climate Change." http://www.dow.com/sustainability/goals/climate.htm (accessed October 2011).
[d]DuPont, "DuPont Climate Change Position Statement." http://www2.dupont.com/Media_Center/en_US/position_statements/global_climate.html (accessed October 2011).
[e]Exxon Mobil, *2010 Corporate Citizenship Report*, at p. 32 (2010).
[f]Ford, "Climate Change Legislation." http://www.corporate.ford.com/microsites/sustainability-report-2010-11/issues-climate-policy-us (accessed October 2011).
[g]Procter & Gamble, "Climate Change." http://www.pg.com/en_US/sustainability/point_of_view/climate_change.shtml (accessed October 2011).
[h]Texas Instruments, "Risks and Opportunities." http://www.ti.com/corp/docs/csr/environment/riskAndOpportunities.shtml (accessed October 2011).

Table 5.3 (Continued)

[illegible]	[illegible]
[illegible]	[illegible] policy options [illegible]
Energy	[illegible]
Lafarge	[illegible]
Holcim [illegible]	We believe that industry, government, and civil society all have roles to play in addressing climate change and that immediate [illegible] action to reduce greenhouse gas (GHG) concentrations in the atmosphere are necessary. [illegible]
[illegible] Insurance [illegible]	[illegible] to provide some much needed perspective on the potential impacts of long-term and regulatory responses.

[illegible]

6 Industry and Multi-State Association Roles

Since 1970, industry trade associations, multi-state coalitions, and professional organizations have emerged as intermediaries between government agencies and members on Clean Air Act compliance. They have forwarded the common interests of their respective interest groups as policy advocates on air quality issues. In turn, these organizations have provided technical assistance to companies on matters of regulatory compliance.

Associations representing automobile manufacturers, chemical companies, the petroleum industry, and other interests have actively participated in policy development for air quality regulations. Intra-industry programs such as the American Chemistry Council's Responsible Care® program have provided a form of self-regulation by requiring proof of external audits from member companies. Trade associations for small-source polluters have offered technical guidance and other services for their constituencies. State air pollution programs in the northeast and other regions coalesced to address area-wide impacts. The Air Pollution Control Association and its successor, American Waste Management Association, have continued to provide key analysis and policy assistance for clean air legislation.

6.1 Associations Representing the Automobile Industry

The Clean Air Act of 1970 placed immediate demands on automobile manufacturers to reduce emissions according to statutory timetables. Through initial years of implementation, competition from fuel-efficient imports cut into domestic vehicle sales and production. One response to these challenges was to establish pre-competitive research and development agreements among domestic manufacturers. The Cooperative Research Act of 1984 provided legislative support for technological innovations among Chrysler, Ford, and General Motors. Joint ventures led to the development of lightweight materials that would improve fuel efficiency. As this effort continued, an Automotive Composites Consortium was formed in 1988. Similar agreements for lightweight batteries and other technologies led to formation of the United States Council for Automotive Research (USCAR) in 1992.[1]

[1] *See* United States Council for Automotive Research, "About USCAR." http://www.uscar.org/guest/history.php (accessed October 2011).

An Interactive History of the Clean Air Act. DOI: 10.1016/B978-0-12-416035-4.00006-0

In 1993, USCAR joined a Partnership for a New Generation of Vehicles with the U.S. government. By 2002, this program transitioned to the FreedomCAR and Fuel Partnership among USCAR, the U.S. Department of Energy, and five energy companies to focus on hydrogen and fuel cell research.[2] This consortium notes that the power of collaboration to increase efficiency, redundancy, and technological innovation at a faster pace than individual company efforts.[3]

> *Advanced batteries enable hybrid electric vehicles that reduce petroleum consumption and gasoline costs. Advanced gas and diesel engines use less gas, perform better and emit less regulated emissions. Advanced lightweighting of vehicles reduces fuel consumption and yet maintains passenger safety.*[4]

USCAR has established additional partnerships with federal agencies, universities, and suppliers.

Chrysler, Ford, and General Motors founded the Automotive Industry Action Group (AIAG). Since 1982, AIAG has supported collaborative efforts in engineering, quality, materials management, and for occupational health and safety.[5] Its membership has expanded significantly to include suppliers and other interested parties. AIAG provides technical assistance, training, and networking opportunities for members[6] The Safety, Health, and Environmental Steering Committee focuses on "chemical issues related to air, water and waste remediation and collaborate with other organizations that can positively influence emerging outcomes and leverage resources."[7]

The Alliance of Automobile Manufacturers superseded the American Automobile Manufacturers Association in 1999 to represent industry interests in environmental, safety, and other matters. It expanded membership to include the BMW Group, Mazda, Mercedes-Benz USA, Mitsubishi, Porsche, Toyota, and Volkswagen. These manufacturers represent 77% of all car and light truck sales in the United States. The Alliance's mission incorporates policy development and implementation for "sustainable mobility":

> *Sustainable mobility focuses on moving people and goods in an affordable and safe manner, while meeting economic, environmental and social goals. To advance sustainable mobility, there is a need for a coordinated effort among automakers, public authorities and other stakeholders.*[8]

This approach encourages further intra-industry collaboration consistent with outreach to governments and other interests.

[2] This partnership includes BP America, Chevron Corporation, ConocoPhillips, Exxon Mobil Corporation, and Shell Hydrogen.

[3] *See* USCAR, "About USCAR." http://www.uscar.org/guest/about (accessed October 2011).

[4] USCAR, *Power of Automotive Collaboration*, at p. 6 (2010).

[5] AIAG, "Membership."http://www.aiag.org/staticcontent/membership/index.cfm?section=membership (accessed October 2011).

[6] AIAG, "About Us." http://www.aiag.org/staticcontent/about/index.cfm?section=aiag (accessed October 2011).

[7] AIAG, *The AAIG Dividend: Creating Supply Chain Value*, at p. 6 (2007).

[8] Global Automotive Industry Meeting, "Leaders of World's Automakers Address Global Issues," January 15, 2005. http://www.jama-english.jp/release/release/2005/050112-01.pdf (accessed October 2011).

The Alliance's "Driving Sustainability" program promotes an integrated approach to reduced emission as well as enhanced energy. It urges energy providers to provide lower carbon fuels, electricity, and supporting infrastructure. The Alliance also seeks consistent long-term governmental policies and consumer support for fuel savings and GHG emissions reductions.[9]

6.2 The American Chemistry Council and the Responsible Care® Program

The American Chemistry Council (ACC) superseded the Chemical Manufacturers Association as the primary trade association for the chemical, plastics, and chlorine industries. These sectors account for 10% of America's export economy. The continuing Responsible Care® program was initiated in 1988.

The Responsible Care® program has been considered effective in extending the industry's stature as a "facilitator of an industry-wide performance and public relations movement."[10] It requires DuPont, Rohm and Haas/Dow, and other member companies to manage pollution prevention along with other health and safety elements. Members must obtain independent certification for compliance with established professional standards. Those standards typically exceed those required by present laws.[11] This organizational structure provides the association a monitoring role within the industry.

The CMA continues to represent industry interests in governmental interactions. It criticized the EPA's requirements that companies modernizing facilities apply for permits when GHG emissions exceed specified levels:

> *EPA's confusing and uncertain permit approval process will curtail new enterprises, significantly reduce investment in the United States and cost jobs. And states that already face budget shortfalls will bear new costs and burdens to process thousands of GHG permits.*[12]

6.3 The American Petroleum Institute

The American Petroleum Institute (API) represents over 400 industry sector members. It works collaboratively within its membership and with related associations "to enhance industry unity and effectiveness in its advocacy."[13] The API's

[9] AIAG, "Why an Integrated Approach Matters." http://drivingsustainability.com/integrated/ (accessed August 2011).

[10] Karen Heller and Ronald Begley, "Redefining the Role and Obligations of an Industry," *Chemistry Week*, July 6–13 (1994).

[11] http://www.americanchemistry.com/s_responsiblecare/doc.asp?CID=1298&DID=5086 (accessed October 2011).

[12] American Chemistry Council, "Environmental Regulations." http://www.americanchemistry.com/Policy/Environment/Environmental-Regulations (accessed October 2011).

[13] API, "Mission." http://www.api.org/aboutapi/mission/index.cfm (accessed October 2011).

environmental principles emphasize the need for science-based risk analysis and overall cost-effectiveness.[14] They also seek development of "cooperative public-private relationships to find lasting, sustainable solutions."[15] Toward this end, API encourages "[p]artnerships within the industry—and with other industries, government agencies and academic institutions."[16]

The National Petrochemical and Refiners Association represents over 450 companies in matters such as the NAAQS, new source review, and climate change policies. One of its primary advocacy issues is to ensure parity among industry sectors in meeting the costs of environmentally based programs.[17] This association is also committed to "continue leading the effort to work with multi-industry coalitions that share concerns over climate change legislation."[18]

6.4 Aerospace Industry Association

The Aerospace Industry Association represents over 300 companies on noncompetitive matters affecting this industry. Its policies integrate environmental issues within a broader category of sustainability, which it defines as "using science to develop new technologies to conserve natural resources." One of the Association's major objectives is to ensure a balance between energy needs and producing equipment capable of operating in extreme conditions.[19] A February 2009 joint statement with 19 other industry-related associations states a commitment to mitigating aviation's contribution to global climate change while continuing industry growth and vitality: "The aviation industry is strongly supportive of continued research to improve scientific understanding of the effects of non-carbon aviation GHGs and the nature of the nitrogen cycle."[20]

[14] API, "Environmental Principles." http://www.api.org/aboutapi/principles/index.cfm (accessed October 2011).

[15] API, "API Public-Private Partnerships Project—Building a Better Future Through Partnerships." http://www.api.org/ehs/partnerships/index.cfm (accessed October 2011).

[16] API, "API Public-Private Partnerships Project."

[17] National Petrochemical and Refiners Association, *Annual Report 2008*, at p. 6 (2008).

[18] National Petrochemical and Refiners Association, *Annual Report 2008*, at p. 12.

[19] Marion C. Blakey, "Viewpoint: An Improving Climate." http://www.aia-aerospace.org/newsroom/publications/aia_eupdate/november_2010_eupdate/viewpoint (accessed October 2011).

[20] Aerospace Industries Association, Air Carrier Association of America, Aircraft Owners and Pilots Association, Air Line Pilots Association, Airport Consultants Council, Airports Council International—North America, Air Traffic Control Association, American Association of Airport Executives, Cargo Airline Association, Experimental Aircraft Association, General Aviation Manufacturers Association, Helicopter Association International, International Air Transport Association, National Agricultural Aviation Association, National Air Carrier Association, National Air Traffic Controllers Association, National Air Transportation Association, National Association of State Aviation Officials, National Business Aviation Association, and Regional Airline Association, "Aviation and Climate Change, The Views of Aviation Industry Stakeholders," at p. 4. Joint Letter, February 2009.

6.5 The Edison Electric Institute

The Edison Electric Institute (EEI) represents the interests of approximately 70% of shareholder-owned utilities and serves 95% of the ultimate customers in the United States. It provides data, analysis, and advocacy for its members with "Congress, government agencies, the financial community and other opinion-leader audiences."[21] In March 2006 testimony on the EPA's proposed $PM_{2.5}$ standard, the Institute criticized the EPA and its science advisors for selecting studies finding health impacts from fine particulates and de-emphasizing studies "that suggest $PM_{2.5}$ presents little or no concern."[22]

In July 2006, the association joined a coalition of industrial trade associations in formal comments on the EPA's proposed rule for implementing the $PM_{2.5}$ standard. These members urged the EPA to extend its schedule for implementing its rules for the $PM_{2.5}$ standard, stating that "the maximum allowable time is necessary to gather air quality monitoring data and develop SIP control strategies that better reflect the implementation of national rules."[23] They recommended further that the EPA should develop options to allow greater regulatory and time flexibility for states with significant nonattainment problems: "[s]uch options should include providing the full 10-year attainment period up front in the 2013 SIP submittal, without requiring a burdensome SIP justification …."[24] EPA's Clean Air Fine Particle Implementation Rule was adopted on March 29, 2007. It requires SIPs to meet EPA's NAAQS for $PM_{2.5}$ by 2010. However, it allows for states to extend its proposed compliance date to 2015.[25] The EEI provides ongoing testimony, filings, and briefs in support of member interests.[26]

6.6 Northeast States for Coordinated Air Use Management

The Northeast States for Coordinated Air Use Management (NESCAUM) offers policy representation and technical support for the common interests of eight states

[21] EEI, "About." http://www.eei.org/whoweare/abouteei/Pages/default.aspx (accessed October 2011).

[22] EEI, "Statement of the Edison Electric Institute, National Ambient Air Quality Standards for Particulate Matter; Proposed Rule," U.S. Environmental Protection Agency Public Hearing, March 8, 2006.

[23] Members of this coalition were: Alliance of Automobile Manufacturers, American Chemistry Council, American Coke and Coal Chemicals Institute, American Forest & Paper Association, American Iron and Steel Institute, API, Corn Refiners Association, Council of Industrial Boiler Owners, Edison Electric Institute, Engine Manufacturers Association, National Association of Manufacturers, National Cotton Council, National Mining Association, National Oilseed Processors Association, National Petrochemical & Refiners Association, National Rural Electric Cooperative Association, Portland Cement Association, U.S. Chamber of Commerce, Utility Air Regulatory Group. "Comments on Behalf of Industry Trade Associations on EPA's Transition to New or Revised Particulate Matter National Ambient Air Quality Standards, Advance Notice of Proposed Rulemaking," July 10, 2006.

[24] Comments on Behalf of Industry Trade Associations on EPA's Transition to New or Revised Particulate Matter National Ambient Air Quality Standards, Advance Notice of Proposed Rulemaking, 71 Fed. Reg. 6718 (February 9, 2006), OAR-2005-0175.

[25] U.S. Environmental Protection Agency, "Final Clean Air Fine Particle Implementation Rule for Implementation of 1997 PM2.5 Standards."

[26] EEI, "Testimony, Advocacy, and Briefs." http://www.eei.org/whatwedo/PublicPolicyAdvocacy/Pages/TestimonyFilingsBriefs.aspx (accessed October 2011).

in air quality and climate programs.[27] It was established in 1967 as a coalition of state pollution control programs concerned with the regional impact of New England power plants. Initial membership included Connecticut, Maine, Massachusetts, New Hampshire, Rhode Island, and Vermont. New York joined in 1970, and New Jersey became a member in 1979. NESCAUM's staff provides technical assistance, scientific and policy analysis to assist its members. Its scope encompasses air quality, climate change, and the effectiveness of regulatory policies.[28]

NESCAUM research and technical assistance links the efforts of Northeast states to protect the New Source Review and other EPA programs, and to strengthen national standards for particulate emissions.[29] Its technical committees integrate resources from member agencies and the EPA regional office.[30] Member states also participate with the EPA in the Northeast Diesel Collaborative. This initiative combines public education, data and analysis, and technology development with "creating new partnerships, programs, regulations, and agreements to reduce emissions."[31] The Northeast Center for a Clean Air Future is a related entity that conducts policy-relevant research that may promote clean air activities by companies and citizen groups.[32]

In referring to these and other ongoing programs, NESCAUM's 2007 annual report encapsulates its overarching role:

> *NESCAUM has helped the Northeast states get things done collectively that they could not easily accomplish on their own. The results of this successful collaboration have long had an influence far beyond the region's borders.*[33]

Its Executive Director described its role as "the 'third leg of the stool' that constitutes our national air pollution control framework ...."[34] The coalition provides an important link between EPA and state/local air agencies. Its cooperative structure has served as a model for similar multi-state forums.

The NESCAUM Climate and Energy Team supports member efforts to reduce GHG emissions, improve energy efficiency, and develop renewable technologies. Another objective is to serve as a forum for addressing these issues. This team is also

[27] Northeast States for Coordinated Air Use Management, *NESCAUM, 1967–2007, Forty Years* (2007). http://www.nescaum.org/about-us/history (accessed August 2011).

[28] *See* Northeast States for Coordinated Air Use Management, "History." http://www.nescaum.org/about-us/history (accessed October 2011).

[29] Northeast Diesel Cooperative, "About the Northeast Diesel Cooperative." http://www.northeastdiesel.org/about.htm (accessed October 2011).

[30] Northeast States for Coordinated Air Use Management, *NESCAUM, 1967–2007, Forty Years*.

[31] Northeast Diesel Cooperative, "About the Northeast Diesel Cooperative." http://www.northeastdiesel.org/about.htm (accessed October 2011).

[32] Northeast States Center for a Clean Air Future, "Overview." http://www.nesccaf.org/about-us/overview (accessed October 2011).

[33] Northeast Diesel Cooperative, "About the Northeast Diesel Cooperative." http://www.northeastdiesel.org/about.htm (accessed October 2011).

[34] Northeast States for Coordinated Air Use Management, *NESCAUM, 1967–2007, Forty Years*, footnote 30, at p. 1.

developing "... credible reporting tools and protocols to assist states, companies, and other entities in quantifying, monitoring, and reporting their GHG emissions."[35]

6.7 State Air Quality Consortiums in Other Regions

Regional initiatives among mid-Atlantic, southeastern, central, and western states provide services comparable to NESCAUM. The Mid-Atlantic Regional Air Management Association represents the common interests of Delaware, the District of Columbia, Maryland, New Jersey, North Carolina, Pennsylvania, Virginia, West Virginia, and local air quality agencies. Its collaborative efforts address common issues in controlling ozone, particulates, and toxic pollutants.[36]

Southeastern States Air Resource Managers include local and state air pollution control agencies from Alabama, Florida, Georgia, Kentucky, Mississippi, North Carolina, South Carolina, and Tennessee. It is dedicated to improving communication within its membership, increasing effectiveness in meeting national and state goals, evaluating air quality issues, and recommending policies and implementation of air quality improvements.[37]

The Central States Air Resources Agencies represent the common interests of Arkansas, Iowa, Kansas, Louisiana, Minnesota, Missouri, Nebraska, Oklahoma, and Texas. It provides a forum for exchange on air quality issues[38] and coordinates policy and technical issues between its members and the EPA.[39] The Western States Air Resources Council was founded in 1988 and has operated since 1992. It currently represents the common interests of 15 states: Alaska, Arizona, California, Colorado, Hawaii, Idaho, Montana, Nevada, New Mexico, North Dakota, Oregon, South Dakota, Utah, Washington, and Wyoming.[40] The Western Regional Air Partnership is a collaborative organization administered by the Western Governors Association and National Tribal Environmental Council.[41] Its primary purpose is to develop technical and policy tools to comply with the EPA's regional haze regulations and other common air quality issues.[42]

[35] NESCAUM, "Climate and Energy." http://www.nescaum.org/focus-areas/climate-and-energy (accessed October 2011).

[36] Mid-Atlantic Regional Air Management Association, "About Us." http://www.marama.org/about-us (accessed October 2011).

[37] Southeastern States Air Resource Managers, "SESARM Purpose." http://www.metro4-sesarm.org/sesarmpurpose.asp (accessed October 2011).

[38] Western Regional Air Partnership, "About the Western Regional Air Partnership." http://www.wrapair.org/about/index.html (accessed October 2011).

[39] Central States Air Resources Agencies, "What Is CENSARA." http://www.censara.org/html/page.php?pageid=17 (accessed October 2011).

[40] Western States Air Resources Council, "What Is WESTAR?" http://www.westar.org/whatis1.html (accessed August 2011).

[41] Western Regional Air Partnership, "About the Western Regional Air Partnership." http://www.wrapair.org/about/index.html (accessed October 2011).

[42] Ibid.

6.8 Associations as Intermediaries Among EPA, States, and Industries

In nearly four decades since passage of the Clean Air Act, trade and multi-state associations have adapted to meet member needs while serving as vital intermediaries with national policy makers. Initial responses challenged the EPA's power to set quality and emission standards. However, their participation has evolved by developing intra-industry environmental standards and monitoring requirements. Industry organizations currently participate in policy advisory groups, partnerships with EPA and states, and as formal representatives in negotiated rulemaking processes.

7 Lessons Learned

From December 1970 forward, the Clean Air Act has redirected administrative structures, operations, and interactive relationships among governments and the industrial sector. The EPA's management of NAAQS, certification of state-administered implementation plans, and regulations for toxic air emissions have reinforced its preeminent role. State governments responded by establishing or expanding programs to attain those standards within federally mandated time frames.

The Clean Air Act also placed unprecedented demands on industries to reduce emissions regardless of costs or currently available technologies. Major corporations and their respective trade organizations escalated environmental affairs to top-level management concerns. Insular and adversary strategies have yielded to emphases on intergovernmental cooperation, early stakeholder involvement, and open scientific exchange. This concluding chapter offers insights on how these public and private sector adaptations may inform future domestic air quality policies. It may also provide insights for other nations committed to protecting public health and welfare.

Since the mid-1950s, the federal role in air quality management has evolved from discretionary assistance to pervasive involvement in matters ranging from indoor pollution to global climate change. Until 1970, administration remained within the domain of local and state administrators. While industries increased attention on environmental concerns, corporate environmental programs had limited authority within their respective management structures. The Clean Air Act reframed those relationships by establishing federal executive powers to issue standards (the NAAQS), guidelines, and certification for state-based implementation programs. The 1990 Clean Air Act amendments placed an additional set of requirements for the EPA, states, and businesses with potential to emit toxic pollutants designated within that law. This extended the net of federal regulation to include businesses such as dry cleaners, print shops, paint dealers, and restaurants.

The EPA consults continually with research and applied scientists to determine the NAAQS. It certifies SIPs and is empowered to issue sanctions that include withholding of transportation funding assistance. This role focuses on overall policy with structured input from industries, trade and multi-state associations, governments, and other interests. The agency partners with states, tribal nations, and industrial representatives on issues ranging from controlling diesel emissions to improving visibility in national parks.

States continue as primary planners and frontline regulators for improving air quality. They administer the Title V federal permit program for toxic pollutants which is required to be self-supporting. State policies and regulatory programs include collaborative implementation and increasing reliance on internal funding sources.

An Interactive History of the Clean Air Act. DOI: 10.1016/B978-0-12-416035-4.00007-2

Industries face continuing challenges to develop new emission-control technologies to comply with federal and state regulations. Corporations have escalated environment compliance matters to top-level executive and board of director concerns. Annual reports reflect these restructuring trends. Companies also report periodically on their initiatives toward sustainable development. Pre-competitive research and development agreements promote technology advancement.

Trade associations represent collective interests in state and federal policy development. They also provide scientific and technical support while advocating industry interests. New trade associations for smaller-scale businesses assist with common regulatory challenges, operating requirements, and informing on required control technologies. Multi-state coalitions offer additional technical analysis to inform policy makers. These entities serve as crucial communication links between regulators and their constituents.

The findings that follow reflect insights from the history of U.S. Clean Air Act implementation. Scientific analysis, technological breakthroughs, and enlightened government policies have emerged through interactions among major actors. Intense debate continues over designating pollutants and their respective standards for public health and safety protection. There are comparably competing interests for determining feasibility, costs, and appropriate timing to implement emission-control technologies. Understanding the interactive lessons from the U.S. experience may expedite effective strategies for other nations confronting the challenges of sustainable development.

1. *Continuing implementation of the Clean Air Act has enhanced the significance of science and technology as underlying bases for public policies.*

- By mid-1971, EPA developed NAAQS for six criteria pollutants: carbon monoxide (CO), ozone (O_3), lead, nitrogen oxides (NO_x), particulate matter (10 or less microns in diameter), and sulfur dioxide (SO_2). The EPA developed NAAQS for fine particulates (PM_{25}) in 1997 and implementation rules for PM_{25} in 2006. It designated GHGs as criteria pollutants in December 2009.
- The EPA bases its NAAQS determinations on Criteria Documents. These documents incorporate epidemiological studies on public health effects of the designated criteria pollutants.
- EPA directed states to regulate mobile source emissions based on California or Federal standards.
  - States such as California and Massachusetts have sought approaches that exceed current national requirements.
  - The corporate case studies in Chapter 5 reference sustainable development and global climate change as priority concerns.
  - Industrial trade organizations focus challenges to the EPA's proposed regulatory standards by critiquing their scientific validity.
- Industries have developed pre-competitive research and development agreements in response to Clean Air Act technological demands.
  - The National Cooperative Research Act of 1984 and National Cooperative Research and Production Act of 1993 encouraged collaborative research to develop technologies that reduce air emissions affecting public health and public welfare.
  - The EPA continues to partner with industries and state government to develop automobile engine and other technologies.

- Debate continues over the causes and appropriate governmental actions to address global climate change. Scientific studies support, refute, and question claims that man-made sources cause atmospheric changes.

2. *Over the course of Clean Air Act implementation, EPA–state–industry partnerships, advisory committees, and negotiated rulemaking have replaced litigation as preferred forums for policy development. Nonetheless, contentious relationships continue over government decisions on particulate regulations, global climate change, and other concerns.*
   - Litigation by environmental groups in the 1970s led to an EPA decision to issue rules directing states not to meet the NAAQS by allowing greater pollution in areas that already meet these standards, and whether states can consider economic factors in administering their implementation plans.
   - Since passage of the 1990 Clean Air Act amendments, the EPA has expanded its reliance on collaborative processes. These include technical advisory panels, negotiated rulemaking, and formal partnerships with industries, tribal governments, and states.
   - EPA regions have expanded cooperative initiatives with industries, state, local, and tribal governments. These include initiatives on diesel fuel emissions and visual quality in natural areas.

3. *Trade associations and interstate alliances have evolved as primary institutions for environmental policy development and technical assistance.*
   - Associations representing major industries have allied for concerted policy advocacy in Washington and in states. States with actively engaged governors and/or committed legislatures have sought policies more stringent than those at the national level.
   - Trade associations for restaurants, dry cleaners, paint retailers, and other smaller-scale businesses emerged to meet gaps in technical guidance and representation in policy forums.

4. *Industrial corporations have prioritized environmental compliance as management concerns. Sustainable development is integral to strategic planning, operations, and to stockholder and government relations.*
   - Each of the corporate case studies in this report reflects significant shifts toward prioritizing environmental compliance strategies.
     - Board-level committees on environmental health and safety communicate directly with CEOs and executive vice presidents.
     - Corporate policies and programs integrate sustainability into policies and practices.
     - Among the case studies, companies with early acceptance of environmental controls gained licensing or distributing technologies to companies and consumers.

5. *Nations in early stages of their pollution control programs may see the benefits in early participation by industry and government representatives.*
   - A 2001 World Bank report on environmental priorities for China cites needs for "building institutional capacity, clarifying administrative responsibilities among national and local agencies, and providing financial support …."[1]
   - China's government has taken a centralized approach to air pollution controls.
     - A joint study conducted for the State Environmental Protection Administration of China and the World Bank (2007) estimated the economic costs from the health

[1]World Bank, *China: Air Land and Water, Environmental Priorities for the New Millennium*, at p. xxiii (2001). http://www.worldbank.org/research/2001/08/1631741/china-air-land-water-environmental-priorities-new-millennium (accessed August 2008).

impacts of air pollution at 157.3 billion yuan, or 1.16% of its GDP.[2] It identified additional costs due to crop loss and forest damages caused by acid rain and SO_2 concentrations.[3]

- China's detailed response to Stockholm Convention on Persistent Organic Pollutants pledged to develop a national strategy for sustainable development. It committed further to "... establish and improve corresponding administrative systems and develop and implement related policies and necessary action measures so as to achieve the control objectives required by the Convention."[4]

• World Bank Environmental Health and Safety Guidelines issued by the International Finance Corporation (April 2007) recommend the following elements for managing air emissions and ambient air quality:

  - *Where possible, facilities and projects should avoid, minimize, and control adverse impacts to human health, safety, and the environment from emissions to air. Where this is not possible, the generation and release of emissions of any type should be managed through a combination of:*
  - *Energy use efficiency*
  - *Process modification*
  - *Selection of fuels or other materials, the processing of which may result in less-polluting emissions*
  - *Application of emission-control techniques.*
  - *The selected prevention and control techniques may include one or more methods of treatment depending on:*
  - *Regulatory requirements*
  - *Significance of the source*
  - *Location of the emitting facility relative to other sources*
  - *Location of sensitive receptors*
  - *Existing ambient air quality, and potential for degradation of the airshed from a proposed project*
  - *Technical feasibility and cost-effectiveness of the available options for prevention, control, and release of emissions.*[5]

• A 2002 analysis of air quality management in the greater Mexico City region recommended incorporating scientific findings into environmental policy while understanding political factors such as the legal system and the effectiveness of government as a negotiator.[6]

• A World Bank support loan to promote environmental sustainability in Mexico identifies institutional and stakeholder risks in: "(i) working with new sectors that have not systematically incorporated environmental considerations into their agendas in the past, and (ii) the promotion of public participation and greater accountability."[7]

[2] The State Environmental Protection Administration of China and The World Bank Rural Development, Natural Resources and Environment Management Unit, East Asia and Pacific Region, *Cost of Pollution in China: Economic Estimates of Physical Damage*, at p. xiii (2007).

[3] Ibid., at pp. 116–20.

[4] The People's Republic of China, *National Implementation Plan for the Stockholm Convention on Persistent Organic Pollutants, Part I*, at p. 1 (April 2007).

[5] World Bank Group International Finance Corporation, *Environmental, Health, and Safety Guidelines, General EHS Guidelines: Environmental Air Emissions and Ambient Air Quality*, at p. 3 (April 2007).

[6] *See* Louisa T. Molina and Mario J. Molina, *Air Quality in the Mexico Megacity*, at p. 31 (2002).

[7] World Bank, *Program Document for a Proposed Environmental Sustainability Development Policy Loan*, at p. 49 (September 5, 2008).

- A 2008 World Bank assessment recommended that India commit to "[a] strong focus on specific desired environmental outcomes and basing project design on a broader up-front stakeholder consultation process ...[.]"[8]

The U.S. experience in air pollution control provides well-established precedent for determining criteria pollutants and NAAQS based on scientific evidence. EPA standards for industrial pollution control technologies are determined with active participation of affected industries. Trade associations, state representatives, and environmental advocates may contest the merits of research methods, causation, or engineering feasibility. However, the *processes* for policy development focus on open exchange to advance the technical basis of regulation. While litigation remains an option for resolving major differences, this alternative can be a latter resort rather than initial strategy to resolve or limit policy conflicts.

[8] World Bank, *Project Performance Assessment Report, India Environmental Management Capacity Building Technical Assistance Project*, at p. 28 [Report No. 44250] (June 23, 2008).

Appendix: State Survey Responses

Project researchers conducted a survey of state programs concerning organizational, fiscal, scientific, and collaborative elements attributable to the Clean Air Act. The summary tables are based on the following survey questions.

1. Is there any documentation that outlines how your state implemented the requirements of the initial Clean Air Act of 1970? Have there been significant changes in the overall organizations responsible for the implementation of the CAA (Clean Air Act) in your state since then?
2. How was this process financed within your state and local governments initially and in subsequent years?
3. What is your state's annual budget related to your state's air quality efforts as a result of the CAA? Do you have an historical account of this funding effort?
4. Who are the key state organizations responsible for the Air Quality Management Plans for your state? What standards of objectivity were applied to evaluate new technologies for SIP development?
5. What are the overall administrative processes used by your state for new source permitting, implementation of best available control technology, and continual revision of the AQMP? This requires development of air quality measurement networks, emission inventories, technology implementation plans, and air quality modeling of both present and future air quality. How is this process implemented within your state? What processes have worked and what has not been as effective?

The project received responses from the following states:

Alabama Department of Environmental Management

State Implementation Plan History	Alabama's State Implementation Plan (SIP) outlines how our state implemented the requirements of the initial Clean Air Act of 1970. There has been one organizational change. Implementation of the CAA was originally the responsibility of the Alabama Air Pollution Control Commission (AAPCC). In the early 1980s, all environmental agencies were combined to form the Alabama Department of Environmental Management (ADEM)
Key state agencies responsible for air quality management plans	The key state organizations responsible for the Air Quality Management Plans in the Alabama Department of Environmental Management (ADEM)
Funding for Air Quality Management Programs	Initially, it was financed by the State's general fund and EPA grant funds. Now, EPA funds, non-Title V permitting fees, and Title V emission fees fund our program. ADEM Air Division's FY00 budget was $8.3 million. Records are available back to 1996
Standards Developed to Evaluate SIP Development	Evaluation of new technologies is done through collaboration and communication of information from EPA and other state and local agencies
Standards developed to evaluate SIP development	We use our Prevention of Significant Deterioration (PSD) regulations, which are based on the EPA's PSD regulations, as a basis for new source permitting. We also routinely adopt new regulations promulgated by EPA that affect sources in Alabama
State administrative processes for new source permitting, BACT, and revision of the AQM plans, and air quality modeling of both present and future air quality	We use our Prevention of Significant Deterioration (PSD) regulations, which are based on the EPA's PSD regulations, as a basis for new source permitting. We also routinely adopt new regulations promulgated by EPA that affect sources in Alabama
Collaborative Participation in Planning and Policy Development	Business Council of Alabama Alabama Environmental Council State and Territorial Air Pollution Program Administrators (STAPPA) Southern States Air Resource Managers (SESARM)

(*Continued*)

Alabama Department of Environmental Management (Continued)

Effectiveness of Collaborative Processes	**1.** We collaborated with the Alabama Petroleum Council and its members to develop regulations to bring cleaner gasoline to our Birmingham nonattainment area since 1998 **2.** We have worked with Alabama Pulp & Paper Council to develop odor regulations for pulp mills

Alaska Department of Environmental Conservation

Agency	
State Implementation Plan History	The resulting State Implementation Plan associated public notices and adopted Administrative Code (regulations). Many of these documents may no longer be available
Key state agencies responsible for air quality management plans	State of Alaska, Department of Environmental Conservation, Municipality of Anchorage, Department of Health and Human Services, Fairbanks North Star Borough
Organizational Changes Attributable to Clean Air Act	The organization has not significantly changed. The Alaska Department of Environmental Conservation was in the mid-1970s (I believe 1974 or 1976). Prior to that the Alaska Department of Health and Social Services handled environmental matters
Funding for Air Quality Management Programs	At state level, the air quality programs have been funded primarily by federal grant and state general fund monies. Since approximately 1995, permit fees have become a significant funding source
Standards developed to evaluate SIP development	In most situations, there are no present standards that apply in all cases. Any proposed regulation must set out the financial consequences of the proposed action. Any emission regulation that is more stringent than an applicable federal requirement is subject to peer review and a need showing based upon health and must consider economic impacts and feasibility. A BACT decision in a permit must meet federal requirements for the analysis which considers economics, energy, and the environment

(Continued)

Alaska Department of Environmental Conservation (Continued)	
State administrative processes for new source permitting, BACT, and revision of the AQM plans, and air quality modeling of both present and future air quality	All administrative regulations to implement new source review permission must go through public review before adoption. Each permit that applies BACT decisions must go through a public review phase. Consideration of all comments is required. Ambient monitoring, emission inventories, etc., are skills, knowledge, and projects done by the state. However, permit applications are required to perform much of this to acquire and maintain their permits
Collaborative Participation in Planning and Policy Development	There are industry trade organizations such as Alaska Oil & Gas association and Alaska Rural Electric Cooperative Association which address common air quality issues. There are local based environmental organizations such as Alaska Center for the Environment, Trustees for Alaska, and others that are actively involved in air quality issues
Effectiveness of Collaborative Processes	We collaborate extensively. Best examples are numerous external workgroups or project panels which are established in support of developing or redesigning an air quality or other environmental program. These panels are usually composed of a very wide array of stakeholders from industry, local government, environmental groups, and tribal organizations. In essence, they create an informal regulation or policy negotiations team; they are not given final decision-making authority, but their recommendations usually carry a heavy preference for the agency decisions

Arizona Department of Environmental Quality	
State Implementation Plan History	Arizona policy for the control of air pollution began in 1962 with legislation authorizing the Arizona Department. of Health Services (ADHS) to conduct air pollution studies and thereby qualify the state for federal grants. In 1967, ADHS was authorized to begin setting air quality standards, and to establish its Division of Air Pollution Control. After passage of the 1970 CAA, additional legislation enhanced the state's role in air quality management.

(Continued)

Arizona Department of Environmental Quality (Continued)	
	Responsibility for regulation of major stationary sources of pollution (copper smelting, power generation, etc.) was assigned to ADHS, and enforcement procedures were established for vehicle emissions. During the 1970s, the emphasis shifted from stationary to vehicular sources. A prototype I&M program was authorized in 1972 and fully implemented in non-attainment areas in 1975. The first SIP was submitted in 1972. In 1986, the Arizona Department. of Environmental Quality was authorized, and all environmental management was moved from ADHS to ADEQ. In 1987, the Omnibus Air Quality Act was passed, enhancing the role of ADEQ in securing improved air quality. Originally, the counties were responsible for most A/Q management, but during the 1980s, all but Maricopa, Pima, and Pinal County ceded authority back to the state.
Key state agencies responsible for air quality management plans	Arizona Department of Environmental Quality (ADEQ), Air Quality Division Maricopa County Environmental Services Department, Air Quality Division Pima County DEQ, Air Quality Program Pinal County Air Quality Control Program Maricopa Association of Governments Pima Association of Governments
Organizational Changes Attributable to Clean Air Act	n/a
Funding for Air Quality Management Programs	Air Quality management activities were initially funded from the state General Fund, which in 1985 amounted to about $600,000, and from federal grants to ADHS, Maricopa County, and Pima County, of about the same amount. In 1974, the Vehicle Emissions Inspection Fund was set up. The Air Quality Fee Fund (AQFF) was established in 1986, and the Air Permit Administration Fund (APAF) in 1992 For the year 2000, the annual budget for the Air Quality Division (AQD) of ADEQ amounted to approximately $14.8 million, compared to perhaps $1 million when these activities were still under ADHS in 1986. Because of several different changes in our accounting systems and organizations over the years, our budget office is not able to provide any more detailed historical information
Standards developed to evaluate SIP development	In general, Arizona statutes and SIP's involving control technologies are derived from Federal mandates, such as New Source Performance Standards and MACT Standards. In some instances, contract research was used to investigate the efficacy and applicability of new control methods. And in some cases, new techniques are keyed off of California benchmarks

(Continued)

Arizona Department of Environmental Quality (Continued)

State administrative processes for new source permitting, BACT, and revision of the AQMP plans, and air quality modeling of both present and future air quality	All rulemaking in Arizona is subject to the Arizona Administrative Procedures Act, which included review by the Governor's Regulatory Review Council. For the nonattainment areas in Maricopa County (the Phoenix region) and Pima County (Tucson region), the Maricopa Association of Governments and the Pima Association of Governments, respectively, have the responsibility. Basically, ADEQ is a pass-through agency for the locally developed SIPs
Collaborative Participation in Planning and Policy Development	Arizona Chamber of Commerce AZ Association of Industries AZ Mining Association AZ Rock Products Association Western States Petroleum Association American Lung Association Sierra Club Arizona Clean & Beautiful
Effectiveness of Collaborative Processes	Part of the process in all our rulemaking activities involves stakeholder meetings to which the appropriate businesses and other interested groups are invited. Other specific examples would be the recent Governor's Brown Cloud Summit and the Agricultural Best Management Practices Committee

California Environmental Protection Agency

State Implementation Plan History	Specific control strategies, compliance schedules, enforcement and monitoring programs, and other state efforts to implement the requirements of the 1970 Act can be found in the 1972 and 1979 California State Implementation pPlans (SIPs). Sections of these two documents have been included for your reference. The documents should be viewed in their historical context. As we have refined air quality control in California, both the governmental structures and form of our programs have evolved.
Organizational Changes Attributable to Clean Air Act	Air quality was first regulated in California at the local level in 1947 when State statute authorized the creation of an air district in every county; Los Angeles established the nation's first Air Pollution Control District later that year. The focus of air quality regulation was initially on stationary sources of visible pollution, such as smoke and particulates. In 1952, our understanding of smog formation improved significantly when Dr. Arie Haagen-Smit discovered that nitrogen oxides (NO_x) and volatile organic compounds (VOCs) form ozone in the presence of sunlight. The California Air Resources Board (ARB or The Board) was created in 1967 by merging the Bureau of Air Sanitation (established in 1955) and the California Motor Vehicle Pollution Control Board (established in 1960) under state law (HSC Sec. 39602). ARB has 11 board members, who serve at the pleasure of the Governor. ARB has about 1,000 staff in 10 divisions and offices.

(*Continued*)

California Environmental Protection Agency (Continued)

	While some local air districts predate the federal Act, many county-level districts were consolidated after 1970 to better address common air quality concerns. In 1976, for example, the South Coast Air Quality Management District (SCAQMD) replaced a voluntary association of air pollution control districts in the Los Angeles region consisting of Los Angeles, Orange, Riverside, and San Bernardino counties. Today, there are 35 local air districts in California. The federal Act requires state to develop SIPs showing how they will meet National Ambient Air Quality Standards within specified time frames. The Act does not dictate the governance structure for meeting these planning requirements, but rather leaves it up to the states to designate the responsible entity. In California, state and local air agencies generally have responsibility for managing different sources of pollution. The state regulated mobile sources, fuels, and consumer products while the locals concentrate on stationary and area sources. Under state law, ARB is the lead agency for implementing the Act. ARB's role in SIP implementation is threefold—to adopt rules that control sources of pollution under its jurisdiction, to review and approve local plans (which focus on reducing emissions from stationary and area sources), and to submit SIPs to the U.S. Environmental Protection Agency (U.S. EPA) for approval The Bureau of Automotive Repair (BAR) and the California Department of Pesticide Regulation (DPR) also have responsibility for emission reductions to meet California's air quality goals. BAR's Smog Check Program was created in 1984 to reduce automobile emissions in specifically designated areas that failed to meet federal and state clean air standards. The legislature has since approved enhancements to the Smog Check program to meet federal law, as well as approving modifications to make the program more consumer-friendly. DPR has authority under state law to control pesticide use (which results in significant VOC and toxic emissions) for the purpose of protecting human health and the environment
Funding for Air Quality Management Programs	The CARB has generally been financed from a mix of state general fund and state special funds. The special funds include the Motor Vehicle Account (funded by vehicle registration fees), the Vehicle Inspection and repair Fund (funded by smog check fees), and the Air Pollution Control Fund (funded by fees levied on vehicle manufacturers and stationary sources of pollution, and penalty assessments). The State also receives annual grant funding from U.S. EPA. The ratio of funding from the General fund and the Motor Vehicle Account has varied over the years depending upon the availability of those funds and the emission source (stationary source activities are primarily funded by the General Fund while mobile source activities are primarily funded by the

(*Continued*)

California Environmental Protection Agency (Continued)

	Motor Vehicle Account). ARB funding mix for fiscal year 2000–2001 is: FY 2000–2001 General Fund 53% Motor Vehicle Account 30% Air Pollution Control Fund 4% Vehicle Inspection & Repair Fund 4% Federal Funds 4% Other Special Funds 5% The source of local air district funding varies from district to district. All districts rely on state subvention funds from ARB (currently $15 million statewide). Larger "grantee" districts also receive federal grants directly from U.S. EPA. SCAQMD, for example, utilize a system of evaluation fees, operating fees, emission fees, hearing board fees, contracts, penalties/settlements, and investments to generate approximately 75% of its revenue. The remaining 25% of its revenue are from U.S. EPA grants, ARB subvention funds, and California CAA motor vehicle fees. The Bay Area Air Quality Management District, on the other hand, utilizes county property taxes to meet some of its funding needs. Smaller (often rural) districts generally are more dependent upon State subvention and federal grant funds
State's annual budget for air quality efforts required by the CAA? Historical account of expended revenue?	The table below describes the Board's operational budget, plus State subvention funds and grants ARB passes on to local districts. These figures do not include other local district funding [...], or funds spent on air quality improvements by the Department of Transportation, BAR, or other State agencies California Air Resources Board Budget Fiscal Year — Amount ($000) 1970–1971 8,131 1971–1972 8,427 1972–1973 13,126 1973–1974 10,065 1974–1975 20,045 1975–1976 23,865 1976–1977 27,065 1977–1978 29,924 1978–1979 38,072 1979–1980 47,912 1980–1981 52,946 1981–1982 56,406 1982–1983 59,538 1983–1984 54,553 1984–1985 47,879 1985–1986 54,691 1986–1987 59,394

(*Continued*)

California Environmental Protection Agency (Continued)

	1987–1988	62,842
	1988–1989	64,542
	1989–1990	78,066
	1990–1991	88,678
	1991–1992	93,128
	1992–1993	100,166
	1993–1994	106,995
	1994–1995	108,558
	1995–1996	115,472
	1996–1997	110,329
	1997–1998	116,748
	1998–1999*	146,095
	1999–2000*	141,149
	2000–2001*	243,087

Key state agencies responsible for air quality management plans	ARB is charged with coordinating efforts to attain and maintain air quality standards, and is the designated lead agency for compliance with federal CAA requirements, including preparation of SIPs. The Board is responsible for controlling emissions from motor vehicles (except when federal law preempts ARB authority), fuels, and consumer products. ARB also assists with technical work, such as emission inventories, air quality modeling, and engineering analysis for local control measures (although some districts conduct their own analyses). ARB consolidates the state and local elements into the SIP, and must review and approve the final SIP before forwarding it to U.S. EPA for federal approval. Once approved by ARB and U.S. EPA, the state and local agencies are responsible by law for ensuring implementation of SIP measures or achieving equivalent emissions reductions Although not part of the SIP, ARB and the local districts work together to reduce emissions and risk from air toxics. ARB establishes the minimum statewide requirements, based on the lowest achievable emission rate, in consideration of risk and cost. Local districts adopt the statewide measure or an equally effective alternative California's BAR and DPR play significant roles in the State's attainment plans—BAR implements the State's Smog Check Program, while DPR manages emissions from pesticides. ARB also works with the Department of Transportation (to develop transportation strategies), the Department of General Services (to ensure a clean vehicle fleet), and the California Energy Commission (to ensure the compatibility of California's energy and clean air goals)

(*Continued*)

California Environmental Protection Agency (Continued)

	ARB coordinates extensively with local air districts, which have primary responsibility to develop the stationary and area source elements of SIPs. Local planning organizations or councils of government (COGs) play a critical role in SIP development by providing the socioeconomic data, transportation and land use information, and growth statistics needed to develop emissions inventories and attainment demonstrations. ARB also works closely with local transportation agencies and COGs to ensure transportation plans "conform" to SIPs, as required by federal law
Standards developed to evaluate SIP development	California has achieved significantly cleaner air over the past three decades by encouraging the development and deployment of new emission-reducing technologies. The three-way catalytic converter, cleaner-burning gasoline, and Zero Emission Vehicles (ZEVs) have all evolved due to ARB programs or regulations. ARB has hosted two new technology symposia to explore emerging technologies to help meet our SIP commitments and identify promising candidates for regulatory development. The Low Emission Vehicle 2 (LEV 2) regulations and 2006 diesel fuel and 2007 diesel engine standards were unveiled at these symposia. ARB has also promoted advanced technologies through its Innovative Clean Air Technologies Program to co-fund the commercialization of technologies to reduce air pollution ARB evaluates new technologies based primarily on three criteria—technical feasibility, cost-effectiveness, and expected emission reductions. ARB examines technical feasibility and control efficiency of potential emission reduction technologies through engineering analysis, pilot programs, testing of prototypes, and demonstrated effectiveness in comparable applications. The Board remains in close contact with manufacturers of emission control equipment to identify promising technologies, enhance our understanding of technical feasibility issues, and evaluate realistic implementation by the scheduled date. California's suggested control measures for architectural coatings, for example, have scheduled reviews to evaluate technological progress and feasibility prior to implementation For criteria pollutants, the cost-effectiveness of a regulation is the estimated total cost of using that technology per ton of emissions reduced. Total cost refers to the total annualized capital costs and operating and maintenance costs (plus research and development and/or installation expenditures, when applicable) of the technology over its lifetime. Annual emission reductions depend on factors such as emissions from the source category before the technology is applied, the technology's expected control efficiency, and its expected market penetration. Cost-effectiveness is particularly important when refining the requirements of a particular rule or guidance

(Continued)

California Environmental Protection Agency (Continued)

	Before a rule or guidance is adopted, ARB staff engages in extensive outreach. This open rule development process, which includes workshops and meetings with the public, environmental and community groups, the affected industry, technology developers, and academia, helps refine the technical evaluations. Throughout this process, stakeholders have access to ARB scientific and technical data, and have the opportunity to present their own research, findings, or perspectives. The Board considers and evaluates all input before adopting a rule or guidance
State administrative processes for new source permitting, BACT, and revision of the AQM plans, and air quality modeling of both present and future air quality	California has a long and successful history of leading the nation in implementing programs to improve air quality. The state's pioneering research on the causes, effects, and methods for control of air pollution provide a strong scientific foundation for these air quality programs. California adopted the nation's first auto tailpipe emission standards for hydrocarbons and carbon monoxide in 1966, followed by the first automobile NO_x emission standards, first use of three-way catalytic converters, limits on lead in gasoline, and vehicle onboard diagnostic requirements. Many of these initiatives have served as models for national programs Research: ARB policies and programs are supported by a strong research program to evaluate how air pollution impacts human health and the environment. ARB conducts extensive research on the impacts of air pollution, with a current focus on children, asthma, and particulate matter. These efforts include the Children's Health Study (to assess the long-term effects of air pollution on lung development in children) and the Childhood Asthma Study (to examine how air pollution impacts childhood asthma) ARB also conducts research to improve and refine its air monitoring, air quality modeling, and emission inventory. Recent field research includes the Central California Ozone Study (to improve understanding of ozone formation and transport across Northern California) and the $27 million California Regional Particulate Matter Air Quality Study (to improve understanding of the formation of gaseous particles and the sources of particulate matter and its precursors) Scientific Foundation: Extensive air quality monitoring networks, emission inventories, and air quality modeling provide the technical foundation for California's programs and regulations. California's monitoring program collects real-time measurements of ambient pollutants at over 40 sites throughout the state. The data generated are used to define the nature and severity of pollution, assess risk, determine which areas are in attainment or nonattainment, identify pollution trends, support agricultural burn forecasting, and develop air models and emission inventories

(*Continued*)

California Environmental Protection Agency (Continued)

ARB has collected information on emissions from air pollution sources since 1969. Criteria pollutant emissions data are compiles on an ongoing basis and stores in the California Emission Inventory Development and Reporting System (CEIDARS). CEIDARS includes emissions from about 13,000 point sources (such as power plants and refineries), hundreds of area-wide sources (such as types of consumer products and residential wood combustion), and mobile sources. Local districts generally compile and report point and area source emissions to ARB. EMFAC2000—a California-specific vehicle emissions estimation model developed by ARB—is used to estimate on-road emissions, while off-road emission estimates are calculated by the OFFROAD model. ARB also collects toxic emissions data from facilities of high risk in California. The toxic pollutant inventory is updated every 4 years and is stored in the Air Toxics Emission Data System

Each region's responsiveness to emission reductions is gauged through photochemical modeling. Modeling is employed to test a wide variety of weather and emission inventory scenarios and assist State and local policy makers develop the most efficient, cost-effective attainment strategies

Mobile Source Program: California is the only state authorized by the federal act to adopt its own motor vehicle emissions or fuel standards and leads the world in advancing new vehicle technology. California's Low Emission Vehicle (LEV) program, adopted in 1990, treats vehicles and fuels as a system required to meet gradually decreasing in-use emission limits and has helped stimulate development of lower-emitting vehicles and cleaner fuels. In 1998, ARB adopted LEV 2 standards to further reduce NO_x and VOC emissions from cars and require light trucks to achieve automobile emission limits. The U.S. EPA adopted more stringent national vehicle emission limits based on LEV 2 in 1999

California's ZEV program, which requires 10% of new vehicles sold in the state by 2003 to emit zero pollution, has stimulated development and commercialization of zero-emission technologies, including fuel cell vehicles. (Manufacturers may meet part of their ZEV requirement by producing other advanced technology vehicles and other strategies.) Section 177 of the 1990 CAA Amendments allows other states to adopt California motor vehicle standards; thus far, Maine, Massachusetts, New York, and Vermont have adopted California's more stringent LEV and ZEV requirements

California is also focusing on diesel engines. New regulations to reduce emissions from on-road diesel engines took effect in 2002–2004 and more stringent standards from off-road diesel engines were phased in. With U.S. EPA, California developed even stricter standards for diesel fuel and on-road diesel engines which took effect in 2006 and 2007. ARB has also adopted fleet rules targeting urban transit buses and implements incentive programs to clean up older diesel engines. These efforts will also reduce emissions of diesel particulate, which ARB identifies as a Toxic Air Contaminant in 1998 (see toxic air contaminant section for further discussion)

(*Continued*)

California Environmental Protection Agency (Continued)

The federal act preempts California from regulating aircraft, new locomotives, international ships, and some types of off-highway farm and construction equipment. As a result, ARB must rely on federal rulemaking to achieve emission reductions needed from these sources to attain national air quality standards. This federal preemption can hinder ARB's ability to pursue the most cost-effective emission reductions, as other sources under state and local jurisdiction are subject to even tighter controls. Although ARB has successfully partnered with U.S. EPA to develop national regulations, emissions from federal sources are increasing and the federal government must do more

Consumer Products: California also has an innovative program to reduce emissions from consumer products like hairspray, deodorants, household cleansers, lighter fluids, and degreasers. Since inception in 1989, California's program has reduced ozone-forming emissions by over 40 tons per day from over 80 categories of consumer products. U.S. EPA has followed California's lead in developing many nationwide consumer product regulations

Stationary Sources: California's stationary source control program is implemented by the local districts. Each district implements its own New Source Review and stationary source permitting programs. Because of the severity of the air pollution problem, many new and innovative stationary source technologies are developed and applied in California. To encourage statewide consistency, ARB has developed Best Available Control Technology (BACT) requirements for power plants, refineries, smelters, and other stationary sources. California facilities generally emit far less pollutants per facility than most other facilities in the nation

California had some difficulty integrating preexisting state and local operating permit programs with the requirements in the Title V of the CAA. By allowing equally effective or more stringent state and local requirements to supersede federal requirements, these programs could operate more efficiently

Toxic Air Contaminants: In 1983, California established a two-phase process for the identification and control of air toxics. In response, ARB has adopted regulations targeting emissions of cancer-causing substances such as benzene, hexavalent chromium, dioxin, and perchlorethylene. Toxic control measures affect over 7,000 facilities, result in reductions in public exposure to these substances by 75–100%, and apply to more facilities than required by national regulations

California's air toxic control measures for dry cleaners, chrome plating, ethylene oxide seltzers, and cooling towers were models for the U.S. EPA's Maximum Achievable Control Technology (MACT) standards. As with the air toxic control measures, California's VOC control measures for sources such as petroleum refineries and gasoline distribution formed the control baseline for the U.S. EPA's MACT standards

(*Continued*)

California Environmental Protection Agency (Continued)

California is also a leader in addressing the toxic particulate emissions from diesel engines. In 1998, ARB identified particulate matter (PM) from diesel-fueled engines as a toxic air contaminant responsible for the majority of potential airborne cancer risk in California. In September 2000, the Board approved a comprehensive Diesel Risk Reduction Plan to reduce diesel PM emissions and health risk by 75% by 2010 and 85% by 2020. The Plan identifies measures to establish more stringent emission limits for new diesel-fueled engines and vehicles, establish particular filter retrofit requirements where technically feasible and cost-effective, and require low-sulfur diesel fuel for on-road and off-road sources

In 1987, California also established the nation's first "Hot Spots" program to address individual facilities that may pose a localized health risk to the public. Facilities must report their toxic emissions, pay emission-based fees, assess risk, and prepare a risk reduction audit and plan (for high-risk sources)

Some provisions of the federal act (or U.S. EPA interpretation of such provisions) have not provided California with the flexibility to effectively pursue its own proven toxic air contaminant control strategies. Implementation of the toxic elements of the 1990 Amendments consumed extensive resources with little health benefit beyond the preexisting state program. California has had particular difficulty implementing its risk-based air toxics program—even though the requirements are likely to be at least as stringent as national standards—because U.S. EPA required a "line by line" equivalency demonstration

Community Health and Environmental Justice: While California's air quality efforts have helped significantly reduce statewide emissions and health risk from air pollution, ARB recognizes the need to specifically address neighborhood-scale air quality issues. Our Community Health and Environmental Justice program seeks to ensure that all individuals in California, especially children, the elderly, and environmental justice communities, can live, work, and play in a healthful environment. The Board is evaluating what air pollution exposures occur in environmental justice communities and seeks to reduce health risk from these exposures as quickly as possible. ARB's Neighborhood Assessment Program will provide the technical tools for assessing cumulative exposure to air pollution and the associated health risk (especially to children) within communities

Compliance: Business assistance programs, run by ARB and local districts, increase compliance with California's air quality requirements. ARB provides technical training courses that keep industry and district enforcement personnel up-to-date on new technology and regulatory changes. ARB also enforces statewide control measures and oversees district enforcement programs for

(*Continued*)

California Environmental Protection Agency (Continued)

	stationary sources of pollution to ensure that emission reduction benefits are achieved and that all businesses are on level playing field Clean Air Plan: Strategies for a Healthy Future: ARB's Clean Air Plan (CAP) will present the Board's long-range vision to achieve California's air quality goals. ARB will conduct a comprehensive assessment of emission reduction opportunities for all sources under State and federal jurisdiction. The result will be potential state and national control measures and emission reduction goals for categories of sources. Selected measures and goals in the approved CAP will then form the basis for new State commitments and federal measures in upcoming SIP revisions The overview perspective afforded by ARB's CAP will integrate ARB efforts to attain health-based standards for criteria pollutants with state initiatives to reduce the public health risk from air toxics. By considering the broad emission reduction needs and opportunities for each source category, the Board hopes to consolidate new control requirements and encourage cost-effective approaches that achieve multiple air quality goals
Collaborative Participation in Planning and Policy Development	Regulated Industry: Virtually every industry in the state—and many out-of-state industries—is affected by California's air quality program. Key industry stakeholders include agricultural interests, oil and gas operators, vehicle and engine manufacturers, and consumer product manufacturers. ARB actively encourages open dialogue with industry before and during the regulatory process, as well as continuing contact afterward to ensure smooth implementation. A sampling of industry groups includes: – Alliance of Automobile Manufacturers – California Cattleman's Association – California Chamber of Commerce – California Council for Environmental and Economic Balance – California Farm Bureau – California Manufacturers Association – California Trucking Association – Consumer Specialty Products Association – Engine Manufacturers Association – National Paint and Coatings Association – Nisei Farms League – Western States Petroleum Association Environmental and Community Groups: Environmental and community organizations have been active in many of ARB's programs, including serving on advisory committees for the

(Continued)

California Environmental Protection Agency (Continued)

implementation of clean gasoline and ZEVs. These organizations are also crucial in public outreach efforts. A sampling includes:

- American Lung Association
- Coalition for Clean Air
- Communities for a Better Environment
- Concerned Citizens of South Central Los Angeles
- Environmental Health Coalition
- Mothers of East Los Angeles
- Natural Resources Defense Council
- Sierra Club
- Southeast Alliance for Environmental Justice
- Union of Concerned Scientists

Government Associations: Governmental associations serve as a forum to improve communication and cooperation among governmental stakeholders, address intra- and interstate air quality concerns, and share technical expertise. A sampling includes:

- California Air Pollution Control Officers Association
- STAPPA/ALAPCO
- Western Regional Air Partnership
- Western States Air Resources Council (WESTAR)

Effectiveness of Collaborative Processes

The scope of California's air pollution problems requires effective collaboration with the regulated community. ARB's rule development process is public by design and provides opportunity to interact with industry both formally and informally. ARB hosts a variety of workshops for the regulated community and routinely consults with business and industry groups. In recent years, ARB and industry have collaborated to develop comments regarding implementation of the federal 8-hour ozone standard, and lobby U.S. EPA for flexibility in implementing the toxics permitting programs called for in the 1990 CAA Amendments

ARB also participates in partnerships to evaluate and promote emerging air pollution control technologies, develop and implement policy, and research the causes, effects, and atmospheric dynamics of air pollution. These efforts include:

- The California Fuel Cell Partnership: ARB is a founding member of this collaboration of auto manufacturers (DaimlerChrysler, Honda, Ford, Hyundai, Toyota, General Motors, Nissan, and Volkswagen), fuel cell developers (Ballard Power, International Fuel Cells), fuel providers (BP, Shell, Texaco), and other government agencies (SCAQMD, U.S. Department of Energy, U.S. Department of Transportation). The Partnership's goals are to demonstrate the viability of fuel cells and an alternative fuel infrastructure, increase public awareness of fuel cells, and explore how to commercialize fuel cell technology. ARB and its industry partners share a new 55,000 square-foot state-of-the-art research, repair, and fueling facility in West Sacramento. The Fuel Cell Partnership aims to have 20 fuel-cell powered transit buses and 50 passenger cars on the road by 2003

(*Continued*)

California Environmental Protection Agency (Continued)	
	• Diesel Advisory Committee: After identifying particulate emissions from diesel-fueled engines (diesel PM) as a toxic air contaminant in 1998, ARB formed the Diesel Advisory Committee to assist in the development of risk management and control strategies. The Advisory Committee (and its fours subcommittees) consists of staff from ARB, U.S. EPA, state and local agencies, industry, environmental groups, and interested public. With the assistance of the Advisory Committee and its subcommittees, ARB developed its Risk Reduction Plan to reduce particulate matter emissions from diesel-fueled engines and vehicles, and risk management guidance for the permitting of new stationary diesel-fueled engines • The California Regional PM Air Quality Study: This multiyear $27 million research project was sponsored and conducted by ARB, oil and agricultural interests, and other governmental and business stakeholders. Results of a field study conducted between December 1999 and January 2001 provided comprehensive information about the origin and effects of particulates and was used to identify a PM control strategy for the Valley • Consumer Products Working Group: The Consumer Products Working Group is an advisory committee that assists ARB in developing and implementing consumer product control measures. The Group includes representatives of industry associations, consumer product manufacturers and formulators, raw materials suppliers, environmental groups, and regulatory agencies

*ARB provided districts with $90 million for the Carl Moyer diesel Retrofit Program between FY 1998–1999 and 2000–2001 and $50 million for the Low-Emission School Bus Program in FY 2000–2001.

Colorado Department of Public Health and Environment—Air Quality Control Commission	
State Implementation Plan History	Colorado implemented the 1970 Clean Air Act requirements by developing a new Air Pollution Control Division. Responsibilities to issue permits, create and enforce regulations, develop plans, monitor the air, and establish mobile source controls have been established by the legislature. Adoption of control plans for all non-attainment areas, complete with local controls, addressed all non-attainment situations. By 1995, all areas of Colorado (17 non-attainment areas) were in compliance with standards.
Organizational Changes Attributable to Clean Air Act	n/a

(*Continued*)

Colorado Department of Public Health and Environment—Air Quality Control Commission (Continued)

Funding for Air Quality Management Programs	Initially, federal funds were the bulk of the Division's funding. Currently, fees from mobile and stationary sources fund 80% of the program
State's annual budget for air quality efforts required by the CAA? Historical account of expended revenue?	The state's annual budget for air quality efforts required by the CAA is $14+ million. We do have historical accounts
Key state agencies responsible for air quality management plans	The Colorado Air Pollution Control Division, the Air Quality Control Commission, nine local health departments, three lead planning agencies, the Colorado legislature, Colorado Department of Transportation
Standards developed to evaluate SIP development	Cost-effectiveness evaluations as part of SID development processes. Also political and social implementability as well as public hearing processes
State administrative processes for new source permitting, BACT, and revision of the AQM plans, and air quality modeling of both present and future air quality	Attainment plans have short-term strategies to establish compliance with standards. When attainment is readied or trends are not showing improvements, a reevaluation of plans is required. During the redesignation/ maintenance plan development, long-term growth and future controls are evaluated to develop an attainment picture for the future. Periodic updates of these attainment plans are required. Maintenance plans contain contingency measures if an area slips back to nonattainment. Evaluation of that option would be assumed to the time when they are needed
Collaborative Participation in Planning and Policy Development	Lead planning agencies Industry stakeholders Environmental groups General public Local agencies
Effectiveness of Collaborative Processes	Subcommittees are usually set up by the AQCC or LPA and are effective collaborators. The division participates in these as need be. Regional Air Quality Control

(Continued)

Florida Department of Environmental Quality	
State Implementation Plan History	Florida's original State Implementation Plan (SIP) submittal to EPA, dated January 27, 1972, is on file, along with all subsequent SIP revisions submitted to EPA.
Key state agencies responsible for air quality management plans	The Division of Air Resource Management (located in Tallahassee), in cooperation with six regulatory districts and eight Approved Local Air Pollution Control Programs
Organizational Changes Attributable to Clean Air Act	There have been no significant changes in organizational responsibility
Funding for Air Quality Management Programs	Initially, the air program process was funded through state general revenue and a 105 AIR Pollution Control Grant from EPA. In subsequent years, this process has been funded through the same 105 Air Pollution Control Pollution control trust fund As a department, the Statewide Air Program is appropriated a $25 million budget. Of the $25 million, the Division of Air Resource Management (DARM) is appropriated an annual budget of approximately $20 million. Of this funding, approximately $10 million is either passed through or contracted out to the eight approved local Air Pollution Control Programs. The remaining $5 million of department funding is allocated to the six regulatory district air quality programs. The total budgets of the Local Approved Programs are unknown. They are funded partly by the state and the rest is through county taxes. We have access to the last 5 years of funding for the state appropriations
Standards developed to evaluate SIP development	Florida Statute 120.541 requires the department to consider the estimated regulatory costs of any proposed new rule and adopt any lower cost regulatory alternative that is presented during the rulemaking process and which subsequently accompanies the objectives of the law

(*Continued*)

Florida Department of Environmental Quality (Continued)

State administrative processes for new source permitting, BACT, and revision of the AQM plans, and air quality modeling of both present and future air quality	Best Available Control Technology (BACT) is a requirement that is triggered by determination that a project is subject to Florida Rule 62-212.400 of the Florida Administrative Code, for the Prevention of Significant Deterioration (PSD) of air quality. The procedures described therein are patterned largely after the general requirements and are incorporated in the federal rules at 40CFR52, Subpart K, Florida State Implementation Plan. Applications for PSD permits include a case-by-case BACT proposal prepared by the applicant. A Professional Licensed Engineer (PE) or an engineer working under the supervision of a PE is required to perform the technology review, cost-effectiveness analysis, and draft BACT determination. The draft BACT is sent out with a sealed technical evaluation and preliminary determination and a public notice. The public and other agencies have 30 days to comment on the draft The proposed permit and draft BACT can be challenged within 14 days by a petition to the department of an administrative hearing which is required by the state's uniform Administrative Procedures Act. Ultimately, the opportunity exists for expert testimony to establish facts and the basis for a final decision (order) with a BACT determination (if permit is issued) The key difference in the procedure with EPA and many other states is that Florida provides for preissuance challenges as opposed to postissuance challenges for EPA permits When a final permit (or order) is prepared, the final BACT determination is included. It is signed by the PE, the Chief of the Bureau of Air Regulation and the director of Air Resource Management for the Department of Environmental Protection The state's air quality management program is largely driven by the requirements of the federal Clean Air Act and associated APE regulations. The DEP Division of Air Resource Management has overall responsibility for keeping track of these requirements and responding to them in a timely manner.

(*Continued*)

Florida Department of Environmental Quality (Continued)

	This is accomplished within the division by organizational subunits having responsibility for the performance or statewide coordination of all activities related to air quality monitoring, emissions monitoring, PSD management, emissions inventories, air quality modeling, state rulemaking, and SIP development. Six DEP district offices and eight DEP-approved local air pollution control agencies have day-to-day operational responsibility for many routine air program functions such as air monitor operations, non-PSD permit processing, compliance inspections, and complaint investigations. Division staff handles the more complex permitting activities, such as PSD permits and Title V permits for utility acid rain units, but mainly function in a planning and coordination role. The consolidation of all air program planning and coordination functions in a single organizational entity provides administrative efficacy Through various administrative mechanisms, the Division of Air Resource Management ensures that the DEP district offices and DEP-approved local air pollution control programs perform all state and federal air management functions as required. Examples of these administrative mechanisms are as follows: General Coordination and Oversight: This includes routine e-mail and telephone communications among division, district and local air program offices, periodic meetings of division/district/local air program managers, coordination of EPA grant air planning agreement commitments between the division and six EPA-funded local air programs, exchange of monthly activity reports, and occasional program audits of district/local air program functions by division staff Conference Calls: To identify air permitting and compliance problems and promote statewide consistency in how they are handled, monthly conference calls are held among both the division/district/local air permitting engineers and division/district/local air compliance staff

(Continued)

Florida Department of Environmental Quality (Continued)

	Guidance Memoranda: To ensure statewide consistency on matters of rule interpretation and the like, guidance memoranda are drafted as needed by the division; circulated among the districts/locals, as well as the regulated community, for review and comment; and published in final form on the Internet
	Specific Operating Agreements: Through specific operating agreements between the division and each of the eight DEP-approved local air pollution control agencies, various air program responsibilities, including compliance inspections, enforcement, and certain kinds of air permitting, are delegated to the local air programs
	Local Program Contracts: Contracts between the division and each of the local air programs are used to transfer funds from the statewide Air Pollution Control Trust Fund to the local programs for statewide Air Pollution Control Trust Fund to the local programs for support of certain air monitoring and Title V permitting and compliance-related activities
	Statewide Database Systems: Statewide database systems for storage and retrieval of air quality monitoring data, emissions inventory data, compliance inspection and test data, and permit tracking data, are maintained by the Division and used by all district and local air program offices
	Statewide Training: The Division coordinated a statewide air training program including sponsorship of air pollution training courses by outside providers; annual workshops for air monitoring, permitting, compliance and emissions inventory staff; training field staff on use of air database systems; and a statewide annual air program meeting for all air program staff
	The division has been fortunate in that we have set up an effective statewide air program. The division, six districts, and the eight approved local programs have been set up to avoid duplication of effort that allows the state to run its air program more efficiently

(*Continued*)

Florida Department of Environmental Quality (Continued)

Collaborative Participation in Planning and Policy Development	The major industry-based organizations that assist the division in our air quality efforts include: Electric Coordinating Group, Pulp and Paper Association, and the Chemical Manufacturers Association. Other groups include the American Lung Association and numerous legal groups that represent other industries such as the municipal waste combustors, sugar industry, etc
Effectiveness of Collaborative Processes	We usually collaborate with the regulated industry prior to rulemaking and legislative processes and it has been an effective process

Hawaii Department of Health

State Implementation Plan History	There is no written documentation as to how the state implemented the initial Federal Clean Air Act of 1970. Pursuant to the Act and through the State department of Health, ambient air quality standards and air pollution control rules were promulgated and became the basis for the air program for Hawaii. Although still within the state Department of Health, the air program has grown from a handful of staff to over 50 personnel with responsibilities consisting of permitting, source monitoring, enforcement, ambient monitoring, laboratory support, and clerical support Hawaii only has a state air program and has no local or county air agencies. Since Hawaii is an island state, one staff each is located on Kauai and Maui, and two are located on the island of Hawaii. The remaining staff is located on Oahu in the Honolulu office
Key state agencies responsible for air quality management plans	The Clean Air Branch of the state Department of Health is responsible for administering and directing the statewide air program
Organizational Changes Attributable to Clean Air Act	n/a

(*Continued*)

Hawaii Department of Health (Continued)

Funding for Air Quality Management Programs	Initially and up to 1997, the air program was predominantly supported with state general funds. As a result of the federal Clean Air Act of 1990, the air program was restructured to be largely supported by fees from the regulated air sources. The air program still receives money from the state general funds and federal grants Historical account of the air funding effort is not available, but the current annual budget for the air program is approximately $5 million
Standards developed to evaluate SIP development	Hawaii is in attainment with the National Ambient Air Quality Standards and therefore has not been confronted with the evaluation of new technologies for the SIP. The evaluation of new technologies has always been contentious for the air program pursuant to the Best Available Control Technology (BACT) determination under the federal Prevention of Significant Deterioration (PSD) Program
State administrative processes for new source permitting, BACT, and revision of the AQM plans, and air quality modeling of both present and future air quality	Hawaii has a network of approximately 18 ambient air quality monitoring stations which provides to the air program an assessment of the air quality throughout the state. In regard to permitting, the sources are categorized into agricultural burning, minor, CAA Title V minor, CAA Title V major, and PSD major sources. The amount and type of information required for an air permit application for a new source or modification are dependent on the site; building dimensions and distances; air quality monitoring data; meteorological date; air modeling with/without surrounding sources; BACT completed, a request for public comments, a public meeting, or a public hearing may be initiated depending on source category or the community sensitivity to the source
Collaborative Participation in Planning and Policy Development	During the rulemaking process, an air advisory committee is convened to assist the air program in the review and comment of the proposed rules. The advisory committee is composed of representatives from the regulated industries, environmental organizations, military, and governmental agencies. Having input from the advisory committee in the early stages of rule development minimizes the adversities and controversies during the public hearing phase of the proposed rules
Effectiveness of Collaborative Processes	The air program does not have any collaborative programs with businesses or industries, although the air program does conduct outreach and educational workshops for specific segments of sources such as dry cleaners and construction activities

(*Continued*)

Idaho Department of Environmental Quality

State Implementation Plan History	The SIP history is documented in the Implementation Plan for the Control of Air Pollution in the State of Idaho, December 1971, amended and reprinted February 1975. This document has been revised numerous times since then. There have been no significant changes in the organization.
Key state agencies responsible for air quality management plans	The state Department of Environmental Quality is the organization responsible for these plans. On local levels there are local organizations that contribute to these plans
Organizational Changes Attributable to Clean Air Act	n/a
Funding for Air Quality Management Programs	EPA 105 funds, state general funds, and special grant funds have traditionally funded the air program in Idaho. When the Title V program began, fees were added to the funding sources. EPA 103 funds were added when the $PM_{2.5}$ monitoring program began The total air budget for SFY2001 is about $4,760,000. This includes EPA 105, 103, and special grant funds, stair air base, and fees. We do not have a historical account of this funding effort.
Standards developed to evaluate SIP development	SIP nonattainment areas look at all new technologies for practicability in applying them to specific nonattainment areas. Economics plays a large role in determining which technologies and strategies get recommended to solve the particular pollution problem being addressed
State administrative processes for new source permitting, BACT, and revision of the AQM plans, and air quality modeling of both present and future air quality	Applications are submitted to the Idaho Department of Environmental Quality (DEQ) by sources. DEQ then determines if the application is complete or not (within 30 days). When an application is determined to be complete, DEQ then has 60 days for the project engineer to write a draft proposal or final permit to construct. For BACT analysis, DEQ uses EPA's Top Down BACT Guidance, the BACT/LAER Clearinghouse, and EPA policy determinations

(*Continued*)

Idaho Department of Environmental Quality (Continued)

	There is no formal process for continual revising of the air program. Policy decisions are made as the need arises in permitting. When we do a rulemaking to revise the current air rules, we usually follow the negotiated rulemaking process. This allows industry, environmental groups, and private citizens to have input on air pollution rules in Idaho Idaho has a series of monitors in various areas statewide. We monitor for PM_{10}, $PM_{2.5}$, CO, Pb, NO_2, and O_3. This network of monitors is maintained by our regional offices and local contractors Emission inventories are completed as needed. Idaho currently has a number of local and statewide inventory projects underway The contracting out of some projects has produced varied results. Sometimes work has to be completed or partially redone by DEQ due to contractor failure to complete the required tasks. We generally get a much better product in house when we have qualifies staff to do the work
Collaborative Participation in Planning and Policy Development	Most organizations get involved with either rulemaking or public comment on air permits. On local levels there are other organizations that are involved in air quality efforts Idaho Cleaners and Laundry Association. Associated General Contractors Idaho Auto Body Craftsmen Association Idaho Association of Commerce and Industry Eastern Idaho Sierra Club Snake River Alliance Idaho Environmental Council Idaho Conservation League
Effectiveness of Collaborative Processes	We do participate in collaborative programs with the business/industries we regulate. For example: Small Business Assistance Program, Pollution Prevention & Environmental Education

(Continued)

	Maryland Department of the Environment
State Implementation Plan History	Organizations responsible for the implementation of the CAA in Maryland have changed significantly since 1970. Initially, air quality control activities took place in a bureau that was part of the Department of Health and Mental Hygiene (DHMH). In 1987, Maryland separated this bureau, which became the Air and Radiation Hygiene (ARMA) from DHMH and formed the Department of the Environment (MDE). The MDE is a regulatory agency and has separate responsibilities from the Department of Natural Resources (DNR), which is responsible for protection of natural resources. Within MDE, ARMA has primary responsibility for implementing the CAA requirements. ARMA has almost 200 employees now, substantially more than were employed in the Bureau of Air Quality Control in the 1970s. ARMA's organization also differs significantly. In earlier decades, ARMA consisted of two programs; one was responsible for permitting and enforcement. ARMA responded to the increasing complexity of air quality control by elevating major areas of the two programs into separate programs so that AMRA now has seven program areas: planning, mobile source control, monitoring and data management, asbestos, radiation, permitting, and compliance.
Key state agencies responsible for air quality management plans	The Air and Radiation Management Administration is responsible for the development of Air Quality Management Plans. As stated above, ARMA coordinated with other state agencies and local governments, but final authority lies with MDE
Organizational Changes Attributable to Clean Air Act	Organizations external to ARMA also changed. Inspection and maintenance programs and stricter transportation conformity requirements fostered a closer relationship between MDE and the Maryland Department of Transportation (MDOT). MDOT provided funding for the inspection/maintenance program as well as other transportation-related programs within ARMA. Eventually, issues surrounding transportation conformity and the development of motor vehicle emissions budgets lead to a fairly close working relationship. The relationship is sometimes strained by pressure to reduce emissions and continue to build roads at the same time, but a compromise has always been reached thus far Under the Intermodal Surface Transportation Act (ISTEA) and the Transportation Efficiency Act for the 21st Century (TEA-21), requirements complementary to the CAA lead to the reorganization of the metropolitan planning organization (MPO) for the Baltimore region. The requirements under the transportation conformity regulation lead to the formation of a subcommittee of the MPO, the Interagency Consultation Group, primarily composed of MDE, MDOT, and the Baltimore MPO.

(*Continued*)

Maryland Department of the Environment (Continued)	
	In the portion of Maryland within Washington, DC, nonattainment areas, a new organizational structure was formed under Section 174 of the CAA. This structure was intended to deal with the problem that not all counties designated as nonattainment in the 1990 Amendments were of the Washington MPO and some county governments preferred to remain within established intercounty liaisons rather than be absorbed by the large metropolitan area. Thus, the Metropolitan Washington Air Quality Committee (MWAQC) was formed to allow local elected official participation in the air quality planning process. This committee is one of the only Section 174 organizations in the country In addition to new local liaisons, several regional processes came to prominence in the 1990s. The MDE has been very active on the Ozone Transport Commission, supporting many of the regional controls proposed through this organization. The MDE was active in the Ozone Transport Assessment Group deliberations. Maryland is part of the Northeast Regional Planning Organization for regional haze. The regional nature of many pollutant and transport issues can only be addressed in these types of forums Over the 30-year period, the State Implementation Plan revisions are the only documents available to show how implementation of the CAA has progressed. Since 1990, the MDE has made a report to the Maryland legislature every 2 years, which is a simpler narrative document
Funding for Air Quality Management Programs	AQM Programs are funded by a combination of state general funds, federal air pollution control program grant and various permits and license fees and penalties (special funds). State general funds provide $800,000 of this total. Historical funding amounts can be ascertained through research
Standards developed to evaluate SIP development	As a small state, Maryland does not evaluate new technologies for use in the SIP. Maryland does evaluate control technology options to set emission rates or other control requirements. In most instances, Maryland works with industry representatives and trade organizations to develop regulations and emission limits. EPA Region III compares regulatory requirements from states within Region III. If one state's regulations are lax, Region III will require revisions to the regulation to make it more stringent. Occasionally, MDE has allowed an industry to install an untried technology with the provision that testing be done to ensure that it achieves the reductions expected

(*Continued*)

Maryland Department of the Environment (Continued)

State administrative processes for new source permitting, BACT, and revision of the AQM plans, and air quality modeling of both present and future air quality	n/a
Collaborative Participation in Planning and Policy Development	The Chamber of Commerce and other business organizations have been active in the regulation development process in Maryland. Various trade organizations have provided technical support and advice on regulations Both the Natural Resources Defense Council and the Environmental Defense Fund have commented on motor vehicle emission budgets, rate of progress plans, and attainment demonstrations. These comments have resulted in changes to the federal transportation conformity rule and in motivating EPA to approve SIPs. These efforts have increased the administrative burden on a state, but have been a positive step for air quality. The Environmental Defense Fund has launched a national commuter choice tax credit program from Maryland
Effectiveness of Collaborative Processes	Maryland businesses and industries have been very active in ENDZONE (now Clean Air Coalition), a voluntary pollution prevention/public outreach program. In this program, Maryland utilities, the American Petroleum Institute, AMOCO, numerous printers, some large banks, and other businesses have provided monetary support or in-kind services to promote pollution prevention activities on high ozone days

Michigan Department of Environmental Quality, Division of Air Quality

State Implementation Plan History	The documentation that outlines how the State of Michigan implemented the requirement of the Clean Air Act of 1970 is contained in the State Implementation Plan (SIP) and subsequent revisions. The initial Michigan SIP, January 1972, addressed the control of suspended particulates, sulfur oxides, carbon monoxide, hydrocarbons, nitrogen oxides, and photochemical oxidants. Numerous revisions to the SIP have occurred since then. Assistance in the preparation of various SIP revisions has been received since then. Assistance in the preparation of various SIP revisions has been received from the Wayne County Air Quality Management Division (WCAQMD), the Southeast Michigan Council of Governments, and the Michigan Department of Transportation.

(*Continued*)

Michigan Department of Environmental Quality, Division of Air Quality (Continued)

Key state agencies responsible for air quality management plans	The MDEO DQ and the WCAQMD are the key state organizations responsible for the Air quality management plans for Michigan
Organizational Changes Attributable to Clean Air Act	Since the initial CAA, there have been no significant changes in the organization responsible for the implementation of the CAA requirements in Michigan, the MDEQAQD
Funding for Air Quality Management Programs	This process was initially financed through federal grant and state general funds. From the mid-1970s to the mid-1980s, the budget was also subsided by surveillance fees. Since 1998, in response to federal CAA requirement, annual air quality fees based on emissions data have supplemented the federal grant and state general funds
Standards developed to evaluate SIP development	The standards of objectivity used in evaluative new technologies for IP development are cost/benefit analysis
State administrative processes for new source permitting, BACT, and revision of the AQM plans, and air quality modeling of both present and future air quality	The state of Michigan's New Source Review (NSR) program is administrated through a central office in Lansing. The Permit Section consists of four units. One unit handles administrative functions, such as scheduled public comment periods and managing the section's permit database. The other three units perform the permit function. Each unit handles specific source categories. This structure was implemented in the late 1980s. Prior to this structure, the two units handled permit review based on geographical area. The change in structure occurred because the geographic setup was too often causing inconsistency in permitting for the same source type Best available control technology (BACT) is implemented in two primary ways in Michigan. For major sources, BACT is implemented pursuant to the federal Prevention of significant Deterioration (PSD) program. Michigan implements a delegated PSD program pursuant to 40 CFR 52.21. For minor sources, BACT is only required for new sources of volatile organic compounds and for sulfur dioxide from natural gas sweetening facilities. State rules approved as part of Michigan's SIP provide the regulatory authority. The rules, as codified in the Michigan Code of Regulation, are R336.1702 (a) and R336.1403 (4), respectively Revisions to Michigan's rules follow standard rulemaking procedures. In addition, there is usually a workgroup of interested parties formed to assist in development of rule packages

(*Continued*)

Michigan Department of Environmental Quality, Division of Air Quality (Continued)	
	Michigan has a state air monitoring network and an emission inventory system. The emission inventory is updated annually, in conjunction with the fee program, for sources subject to Title V of the federal CAA. The air monitoring program complements the federally mandated program. The AQD has a unit that does all of the dispersion modeling required for permit reviews, emission inventory, and SIP work Michigan has implemented several processes in the Permit Section that have improved the NSR process. The most important of these was to establish a position to screen incoming permit applications for administrative completeness. Some 80–90% of permit applications submitted to the state are incomplete in some way. The screening process is very effective. It ensures that very incomplete applications are corrected before being assigned to a permit reviewer. It also provides an initial notification to the applicant that the application was received and meets basic requirements for further processing Another process is the use of General Permits. Michigan has issued five General Permits. Each permit covers a source category as concrete crushers, ethylene oxide sterilizers, or small remediations. The General Permits are limited to small sources that meet certain requirements
Collaborative Participation in Planning and Policy Development	The major organizations that assist in our efforts include: American Automobile Manufacturers Association American Lung Association Associated Petroleum Industries Michigan Manufacturers Association Michigan Chemical Council Michigan Environmental Council State and Territorial Air Pollution Program Administrators
Effectiveness of Collaborative Processes	Examples of collaborative programs with businesses and industries that we regulate include: NO_x SIP Development Southeast Michigan Ozone Study Lake Michigan Air Directors Consortium Development of attainment demonstrations/maintenance plans Various internal and external advisory groups

(Continued)

Mississippi Department of Environmental Quality

State Implementation Plan History	Air Emissions Regulations, APC-S-1, Construction and Operating Permit Regulations, APC-S-2
Key state agencies responsible for air quality management plans	Mississippi Department of Environmental Quality
Organizational Changes Attributable to Clean Air Act	n/a
Funding for Air Quality Management Programs	Initially with state general funds and federal grants. In subsequent years, we have added Title V fees to the mix
State's annual budget for air quality efforts required by the CAA? Historical account of expended revenue?	Approximately $5.5 million
Standards developed to evaluate SIP development	
State administrative processes for new source permitting, BACT, and revision of the AQM plans, and air quality modeling of both present and future air quality	See Regulations APC-S-2 and Title V Permit Regulations, APC-S-6
Collaborative Participation in Planning and Policy Development	Mississippi Manufacturers Association, Mississippi Economic Council
Effectiveness of Collaborative Processes	We do collaborate in certain programs. One example is the Gulf Coast Ozone Study

Missouri Department of Natural Resources Air Pollution Control Program

State Implementation Plan History	There are annual reports dating back to 1972 that show budget, staffing, and duties.
Key state agencies responsible for air quality management plans	The mission of the Department of Natural Resources' Air Pollution Control Program is "to maintain purity of the air resources of the state to protect the health, general welfare, and physical property of the people, maximum employment, and the full industrial development of the state." The program serves the public with technology, planning, enforcement, permitting, financial, and information services to achieve this mission Technical Support: The department's program staff analyzes the quality of Missouri air using chemistry, meteorology, mathematics, and computer programming. Staff members research the sources and effects of air pollution, collecting and maintaining an annual inventory of sources that emit air pollution. In conjunction with the Department of Natural Resources' Environmental Services Program and four local agencies, the department's Air Pollution Control Program staff designs and coordinated an air-monitoring network

(Continued)

Missouri Department of Natural Resources Air Pollution Control Program (Continued)

and analyzes monitoring data. The network provides air quality data from more than 40 locations around the state. Using the monitoring data and other data on source emissions and the weather, the staff runs computer models of the atmosphere to predict air quality

Planning: The department's program staff develops rules and plans designed to protect Missouri's air quality. Public participation is a vital part of the cooperative process of developing guidelines and regulations. The staff works with businesses, federal, state, and local government agencies, environmental groups, and the public in a number of ways, including exchanging ideas and information on clean air issues with advisory groups, workgroups, and in workshops

The department's program staff works closely with the U.S. EPA as part of the national effort to improve the air quality through the Clean Air Act. The staff researched and analyzes complex environmental issues to develop air pollution control strategies that will ensure Missouri's progress in achieving and maintaining air quality improvements. These air pollution control strategies are included in state implementation plans to control specific pollutants. The Missouri Air Conservation Commission approves the state implementation plans and rule actions after they have gone through a public hearing process. Once rules are adopted by the Missouri Air Conservation Commission, they become effective through publication in the Missouri State Code. State implementation plans and associated rules adopted by the Missouri Air Conservation Commission are submitted to the U.S. EPA for inclusion in the federally approved state plan

Permits: The department's program staff review construction permit applications of new and modified emission sources to ensure that facilities minimize the release of air contaminants and will meet all laws and regulations requirements. Operating permit applications, similar to business licenses, are also received and reviewed. Operating permits identify all the air pollution control requirements of a source of air pollution

Enforcement: The department's program staff responds to complaints about air quality and helps businesses comply with various federal, state, and local rules. Staff conducts routine site inspections and oversees the testing of smokestacks, asbestos removal, gasoline vapor recovery equipment, and other sources of air pollution through regional offices. When a source violates an air quality requirement, the staff works with the facility to correct the problem and may take additional action, including the assessment of penalties necessary to obtain compliance with the requirement. Cases that cannot be resolved are referred to the attorney general's office through the Missouri Air Conservation Commission

(*Continued*)

Missouri Department of Natural Resources Air Pollution Control Program (Continued)

	Administration: The department's program staff provides budgeting, procurement, public information, and personnel services. The staff also provides liaisons for the Missouri Air Conservation Commission, the U.S. EPA, the Missouri Department of Health, local air agencies in St. Louis County and cities of Kansas City, St. Louis, and Springfield, the American Lung Association, and the news media. Local Agencies: Four local governments in Missouri practice the regional control over air pollution: Kansas City, St. Louis County, and Springfield. A city or county may have its own air agency under two conditions: The city must be able to enforce its rules and its rules must be as strict as the state's. Local agencies issue permits, maintain their own monitoring networks, and may enforce asbestos removal laws. The local agencies are partially funded by the U.S. EPA through the Department of Natural Resources
Standards developed to evaluate SIP development	The Air Pollution Control Program recently revised the control strategy for the Herculaneum Lead SIP. The SIP submittal involved the development of an emission inventory protocol, observation of emission testing, oversight and review of on-site meteorological data, development of a comprehensive hour-by-hour emission inventory, the development and considerable refinements of a dispersion model, three rounds of receptor modeling, and model reconciliation. The emission control strategy involves enclosure of the main process at the plant and the installation of building ventilation systems. The ventilation gases will be filtered by state-of-the-art, high-efficiency baghouse filtration systems prior to release to the atmosphere. Capital costs are expected to be about $12,000,000. As part of the SIP development, the EPA strongly recommended using a different modeling tool. Chemical Mass Balance modeling is a statistical method of quantifying individual source contributions by examining the chemical profile or "fingerprint" of each source and comparing this to samples collected in the ambient environment
State administrative processes for new source permitting, BACT, and revision of the AQM plans, and air quality modeling of both present and future air quality	Issuance of a Best Available Control Technology (BACT) is the result of a major permitting effort and is similar to adopting a regulation nationwide. When the state sets an emission limit, new sources that construct anywhere in the nation after that time must operate within that limit or justify why they cannot. Review of Archer-Daniels-Midland's soybean oil extraction cycle turbines in southeastern Missouri and the Kansas City Power and Light's coal-fired power plant in Kansas City set the standard for businesses nationally. Important BACT analyses resulted from the review of these projects. Permitting of these facilities in Missouri raised the bar for similar facilities across the country

(*Continued*)

Missouri Department of Natural Resources Air Pollution Control Program (Continued)

Collaborative Participation in Planning and Policy Development	The department actively participates in air quality meetings of the two major metropolitan planning organizations, East-West Gateway Coordinating Council in St. Louis and Mid-America Regional Council. At these public meetings, the department provides updates on air quality projects and discusses proposed rules and plans with other participants The department also relies on the Small Business Compliance Advisory Committee. The three components are a small business ombudsman, a technical assistance program for small business and a compliance advisory panel. It assists small businesses that are often focused on their day-to-day operations and may find it difficult to keep up with changing air pollution regulations and requirements. The Small Business Compliance Advisory Committee has the following responsibilities: • Receives reports from the small business ombudsman (governor's office) • Evaluates the impact on small business of the Air Conservation Law and related regulations • Makes recommendations to the Missouri Department of Natural Resources, the Missouri Air Conservation Commission and the General Assembly regarding changes in procedure, rule or law that would help small businesses comply with the Air Conservation Law • Makes recommendations to the Missouri Air Conservation Commission on rules to expedite the review of modifications for small business • Conducts hearings and make investigations consistent with the purposes of the small business technical assistance activities
Effectiveness of Collaborative Processes	Involving the public in the process of making air quality rules to create fair, effective regulations that have broad support. In 1999, DNR continued its commitment to public participation by convening workgroups to help develop air regulations. A workgroup brings industry and the public together with government agencies to share concerns and exchange ideas while developing regulations The Construction Permit Streamlining Workgroup continued improving the Construction Permit Regulations and reviewing the internal procedures and policy for the program to review permit applications. After receiving final recommendations, the Missouri Air Conservation Commission adopted the proposed amendment to the construction permit rule on July 29, 1999

(*Continued*)

Missouri Department of Natural Resources Air Pollution Control Program (Continued)	
	The department also worked with leaders from industry, environmental organizations, and local government to improve air quality in the Kansas City area. The department participated as a member of the Mid-America regional Council, a metropolitan planning organization, in the development of an air quality improvement plan for the Kansas City ozone maintenance area, which includes Johnson and Wyandotte counties in Kansas and Clay, Jackson and Platte counties in Missouri. In June 2000, DNR participated in the Kansas City Fuels Summit. Discussion focused on determining a motor vehicle fuel strategy to improve air quality in the Kansas City ozone maintenance area

Montana Department of Environmental Quality	
State Implementation Plan History	Administrative rules from the 1970s are available from the Montana Secretary of State. The major change in the overall organizations responsible for the implementation of the CAA in Montana would be the creation of the Department of Environmental Quality (DEQ) in 1995. This brought all environmental regulatory programs together under one agency. The DEQ was structured along functional lines to integrate environmental regulation of the various media (air, water, land).
Key state agencies responsible for air quality management plans	Statutory authority to implement the Montana Clean Air Act (MCAA) was delegated to the Department of Health and Environmental Sciences (now DEQ) in 1967. Therefore, the Montana DEQ is the key state organization responsible for air quality management plans. However, Part 3 of MCAA allows for the formation of local air pollution control programs with the approval from the Montana Board of Environmental Review. Montana currently has seven county air pollution control programs. These local programs are also key organizations responsible for the formation of air pollution control plans
Organizational Changes Attributable to Clean Air Act	n/a
Funding for Air Quality Management Programs	Initially, the air quality program in the State of Montana was funded through EPA grant funds and state general funds. With the passage of the Clean Air Act Amendments of 1990, the state began assessment of air quality permit fees and phased out use of general fund dollars. The program is currently funded through EPA grant funds and fees Since the Montana Department of Environmental Quality is structured along functional lines and not media line[s], it is not possible to accurately break an air quality budget. Information for the previous department is not available at this time

(Continued)

Montana Department of Environmental Quality (Continued)

Standards developed to evaluate SIP development	Development of particulate matter, carbon monoxide, sulfur dioxide, and lead control plans in Montana have followed EPA's RACM/BACM or RACT/BACT requirements as set forth in Sections 189 and 190 of the Clean Air Act. Standards of objectivity have resulted in the identification of source contributors and the development of control measures that serve to regulate only the extent of a source contribution. Alternative strategies involving new technologies have not been incorporated into Montana SIP development. This is, in part, due to low emission control credits allowed by EPA
State administrative processes for new source permitting, BACT, and revision of the AQM plans, and air quality modeling of both present and future air quality	BACT is required for every emitting unit required to obtain an air quality permit. This would apply to each facility with a potential to emit of 25 tons per year or more
Collaborative Participation in Planning and Policy Development	Montana participates in STAPPA (State Air Pollution Control Administrators Association), WESTAR (Western States Air resources Council), [and] The Western Governors Conference. The state also confers with in-state industrial organizations and environmental organizations
Effectiveness of Collaborative Processes	The department of Environmental Quality has established an informal advisory group called the Clean Air Act Advisory Committee to advise the department on rulemaking, legislative, fee, and budgetary issues. This group is composed of regulated industries, environmental groups, and other interested parties

Nebraska Department of Environmental Quality

State Implementation Plan History	State Implementation Plans; Rules & Requirement
Key state agencies responsible for air quality management plans	Department of Environmental Quality has responsibility for overall State Implementation Plan. Three local agencies have partial responsibility: Lincoln/Lancaster County Health Department Omaha Air Quality Control Douglas County Health Department
Organizational Changes Attributable to Clean Air Act	Changes to the organization include growth in number of staff, opening of field offices, restructuring of organization, and collaboration with local agencies

(Continued)

Nebraska Department of Environmental Quality (Continued)

Funding for Air Quality Management Programs	Funding has been pretty typical: through state general funds, federal grants, local general funds, and most recently Title V emission fees 2000 Budget Title V—$1,900,000; 105 federal grant/ general fund—$1,325,000; 103 federal grant—$390,000
Standards developed to evaluate SIP development	Nebraska does not have any nonattainment areas where this would be applicable
State administrative processes for new source permitting, BACT, and revision of the AQM plans, and air quality modeling of both present and future air quality	We have a state construction program and are a delegated agency for PSD. We have a broad ambient monitoring network; require annual emission inventories from most "significant" sources and conduct modeling for most permitting actions
Collaborative Participation in Planning and Policy Development	NICE; AWMA; STAPPA/ALAPCO; CENSARA; Neb Health 7 Human Services, Nebraska Department of Economic Development; and various industry-specific groups; CITIZENS; CCAACA
Effectiveness of Collaborative Processes	We have a partnership agreement with Nebraska Public Power District

New York State Department of Environmental Conservation

State Implementation Plan History	The primary documentation that outlines how New York State implemented the CAA would be State Implementation Plans (SIPs). We have copies of SIPs back to 1974. New York State has always been responsible for the implementation of the CAA in the State
Key state agencies responsible for air quality management plans	New York State Department of Environmental Conservation New York State Department of Health New York State Department of Transportation
Organizational Changes Attributable to Clean Air Act	n/a
Funding for Air Quality Management Programs	New York State's annual budget related to the State's air quality efforts as a result of the CAA was over $42 million for State Fiscal Year 2001/02. Federal funding provided approximately $9 million of this amount with State funding providing the remainder

(Continued)

New York State Department of Environmental Conservation (Continued)

	We have historical records back to State Fiscal Year 1999/2000 (April 1, 1999 to March 31, 2000)
Standards developed to evaluate SIP development	New technologies for SIP development are evaluated using Guideline Emission Models, Affirmed AP-42 emission Factors, or case-specific test data
State administrative processes for new source permitting, BACT, and revision of the AQM plans, and air quality modeling of both present and future air quality	Please call John Higgins, Chief, Bureau of Stationary Sources of the New York State Department of Environmental Conservation, Division of Air Resources at (518) 402–8403 for a detailed response to this question
Collaborative Participation in Planning and Policy Development	State and Territorial Air Pollution Program Administrators/Association of Local Air Pollution Control Officials (STAPPA/ALAPCO) Northeast States for coordinated Air Use Management (NESCAUM) Ozone Transport Assessment Group (OTAG) Ozone Transport Commission (OTC)
Effectiveness of Collaborative Processes	Four examples of our collaborative efforts with regulated businesses/industries are: **1.** New York state was involved in a joint testing project with New York City Transit/Metropolitan Transit Authority (NYCT/MTA) and various private corporations to test a technology that will reduce diesel emissions on heavy-duty diesel buses **2.** Dry cleaning associations were represented on the committee that developed rules regarding perchlorethylene emissions generated form that industry. There are also three dry cleaning associations on an advisory board that was formed to discuss issues relating to these rules with the Department

(*Continued*)

New York State Department of Environmental Conservation (Continued)

	3. During the promulgation of the Stage 1 and Stage 2 Vapor recovery regulations, numerous meeting and seminars were held with the major oil companies, equipment manufacturers, service station owners, and the trucking industry to work out questions about equipment compatibility at the terminal loading racks and service stations. The purpose of the meeting and seminars was to ensure that all gasoline delivery lines and connections, vapor recovery lines and connections, and the safety and spill prevention hook-ups would be compatible statewide **4.** During the promulgation of the Heavy-Duty Gasoline Truck Inspection/Maintenance regulation, an economic impact assessment of the regulation on the downstate/Metro-New York City trucking industry was undertaken. The results of this collaborative effort led to a regulation that allowed fleets to self-inspect and helped the industry prepare and implement regular programs to better maintain their trucks which reduced the probability that they would fail the emissions test and require additional downtime to repair their trucks

Oklahoma Department of Environmental Quality

State Implementation Plan History	Our original State Implementation Plan was submitted to the EPA on January 28, 1972. We still have a copy of that submittal, and so should EPA Region 6. We also have audiotapes of our Air Quality Council proceedings dating back to 1971. The Oklahoman Department of Health was the organization responsible for CAA implementation in Oklahoma. That function became the responsibility of the newly created Department of Environmental Quality in 1993.
Key state agencies responsible for air quality management plans	The Oklahoma Department of Environmental Quality is the primary agency for developing and implementing the air pollution control program. The automotive anti-tampering inspection program that is part of our State Implementation Plan is administered by the Oklahoma Department of Public Safety

(*Continued*)

Oklahoma Department of Environmental Quality (Continued)

Organizational Changes Attributable to Clean Air Act	n/a
Funding for Air Quality Management Programs	The process was initially funded through the EPA's grant funds and state appropriations. Currently funds come from state appropriations, EPA grant funds, and annual operating fees and other fees for services rendered Recent annual budgets: \$7.38 million FY 2001 \$6.92 million FY 2000 \$5.83 million FY 1999 Historical data not available
Standards developed to evaluate SIP development	We generally rely on guidance from EPA, the State and Territorial Program Managers Association, and private consultants in evaluating the effectiveness of new control technologies
State administrative processes for new source permitting, BACT, and revision of the AQM plans, and air quality modeling of both present and future air quality	Administrative processes used for new source permitting implementation of BACT include training for permit writers in the use of the EPA New Source Review Workshop Manual which clearly outlines procedures and practices for BACT determination, use of the EPA database RACT/BACT/LAER Clearinghouse, consultation with other states concerning latest technology and monthly conference calls with the Permit Section of the EPA Region VI office. AQD staff also attends ongoing training and workshops supplied by EPA, STAPPA-ALAPCO, and CenSARA in issues that involve BACT. BACT determinations within the Oklahoma Air Quality Permits Section are shared weekly during the supervisor meetings and then passed on the individual permit writers for their use Administrative procedures used for continual revision of the State's Air quality Management Program include: Air Quality Modeling: Additional Urban Airshed Modeling has been required by the DEQ for newly proposed electric power plants to determine the overall effect of these facilities on Oklahoma Air Quality Technology Implementation Plans and Emission Inventories: The annual emission inventories database is being combined with the Air Quality TEAM database that holds the individual permits and compliance and enforcement information in a single database

(Continued)

Oklahoma Department of Environmental Quality (Continued)	
	Air Quality Measurement Networks: Acid rain permits have emissions monitoring requirements; Title V permits require strict record-keeping and measurement of emissions; the air Quality Division has a statewide monitoring network that is continually improved and updated; and Air Quality maintains some temporary regional (portable) monitors to investigate the quality of the air entering Oklahoma. Data from future and/or present tribal monitors will be considered and evaluated by the Air Quality Division
Collaborative Participation in Planning and Policy Development	Environmental Council of the States State and Territorial Air Program Administrators Central States Air Resources Agency Mid-Continent Oil and Gas Association Citizens Action for a Safe Environment Oklahoma Grain and Feed Association Environmental Federation of Oklahoma Associations of Central Oklahoma Governments Indian Nation Council of Governments Association of Southern Oklahoma Governments
Effectiveness of Collaborative Processes	As part of the rulemaking process, the Air Quality Division of DEQ often establishes workgroups composed of industry representatives, division staff, and the interested public. As an example, the agricultural community plays a significant role in Oklahoma. While in the process of revising out regulation dealing with the Control of Emissions of Grain, Feed, and Seed Operations, workgroups with significant industry representation proved to be the most expeditious ways to reach common goals in the rulemaking process

Rhode Island Department of Environmental Management	
State Implementation Plan History	There is no documentation about the implementation of the 1970 CAA. The only change in organization is that environmental programs, including the responsibility for clean air, moved from the Department of Health to a newly formed Department of Environmental Management in 1977.
Key state agencies responsible for air quality management plans	The Department of Environmental Management has statewide responsibility for Air Quality Management Plans

(Continued)

Rhode Island Department of Environmental Management (Continued)

Organizational Changes Attributable to Clean Air Act	n/a
Funding for Air Quality Management Programs	All implementation of the Clean Air Act in Rhode Island is done at the state level. Funding has, until recently, predominantly been from the federal air pollution control grant under section 105 of the Clean Air Act. State appropriations have provided funds to match the federal grant, traditionally near the minimum match requirement. State funding increased to the 40% match level when required. Title V fees now provide about 25% of program funding
Standards developed to evaluate SIP development	Rhode Island normally does not "evaluate new technologies" in SIP development. Rather, SIP development is based on adoption of control programs that are in place or in process in other states in the Northeast, California, or the remainder of the country
State administrative processes for new source permitting, BACT, and revision of the AQM plans, and air quality modeling of both present and future air quality	Rhode Island operates a delegated permitting program for NSR and PSD. We also require permits for new and modified sources. The basic permit requirement is the application of BACT for minor sources down to a threshold of emission of 10 pounds per hour or 100 pounds per day for any pollutant There isn't a formal process for continual revision of Rhode Island's air quality program. Your model above left out enforcement as well as outreach and compliance assistance. Rhode Island is part of the evolution of environmental regulatory programs toward compliance assistance and away from strict enforcement. Our Department is also seeking a more multimedia approach
Collaborative Participation in Planning and Policy Development	We occasionally work with an organization of plant managers, but do not have an ongoing relationship with an industry-based organization (but see below). We work with the American Lung Association of Rhode Island
Effectiveness of Collaborative Processes	The Department has an ongoing Business Roundtable and an Environmental. The purpose is to inform, discuss, and take suggestions from each group on whatever are topical issues

(*Continued*)

Texas Natural Resource Conservation Commission

State Implementation Plan History	The initial implementation of the Clean Air Act of 1970 was through Chapter 382 of the Texas Health and Safety Code, also known as the Texas Clean Air Act, which was originally published in 1965. However, the first air quality initiative in Texas was established in 1956, when the State Department of Health, Division of Occupational Health and Radiation Control, began air sampling in the state. The first State Implementation Plan to improve Texas air quality was published in 1972. The organization responsible for implementing the CAA has changed significantly over the years from a service within the Texas Health Department, to an independent Air Control Board, to a portion of a Natural Resource Conservation Commission.
Key state agencies responsible for air quality management plans	The key state organization responsible for Air Quality Management Plans in Texas is the Texas Natural Resource Conservation Commission
Funding for Air Quality Management Programs	Initially, all activities were funded through general revenues with some federal grant funds added later. In about 1984, Texas State Air Pollution Control Activities became partially funded through permitting fees. In 1985, inspection fees were added to help finance the process. About 1992, emission fees and vehicle inspection fees were used to totally finance implementation of the CAA (no general revenue) The state budget for the organizations which implement air quality is contained in the biennial Texas State Appropriations Act, Clean Air Account 151. Information on current budget expenditures is also contained in Chapter IV of Report to the Texas Sunset Advisory Commission, August 1999
Standards developed to evaluate SIP development	Control technologies proposed for the SIP have always considered technical feasibility. To support this review, the agency developed central control measure catalogs which considered rule effectiveness, rule penetration, affected parties, estimated costs, and cost-effectiveness
State administrative processes for new source permitting, BACT, and revision of the AQM plans, and air quality modeling of both present and future air quality	The Air Permits division (about 200 positions) deals with permitting and authorization of all air emission sources within the state. Texas addresses every man-made source of air pollution, no matter how small. The permitting process applies BACT, at a minimum, to every source which applies for a permit. The permitting flow process is described on pages 201 through 210 of the Report to the Sunset Advisory Commission.

(Continued)

Texas Natural Resource Conservation Commission (Continued)

The SIP process impacts both permitted and nonpermitted sources of air emissions in Texas by making changes to the air quality rules. The Technical Analysis Division (about 300 positions) includes a Modeling and Data Analysis Section which conducts airshed modeling to demonstrate the reductions to specific source category emissions within the airshed resulting from changes to air rules to allow the area to meet national air quality standards. Areas and Mobile Emissions Assessment Section develops criteria emission estimates for area and mobile sources using EPA-approved mobile source models and emission factors. The Industrial Emissions Assessment Section conducts an annual emissions inventory of point source emissions for SIP analysis and development. The Monitoring Operations Division (about 120 positions) deploys and maintains more than 120 air monitoring stations in Texas which monitor ambient levels of criteria pollutants and provide feedback to the state and federal governments on attainment status of national air quality standards. One example of how this process is implemented within Texas is the establishment of a steering committee to develop local control strategy options to augment federal and state programs in the Dallas–Ft. Worth area. The steering committee made up of local elected officials and business leaders identified specific control strategies for review by technical subcommittee members. In addition, the North Central Texas Council of Governments hired an environmental consultant to assist with the analysis and evaluation of the various options. The consultant was responsible for presenting the findings of the technical subcommittees to the steering committees for final approval prior to being submitted to the state. The total package of local, state, and federal control options now form a compelling argument that attainment can be achieved in the Dallas–Ft. Worth area

The SIP process does work and continues to improve, but it is also technically and functionally very complex and resource-intensive. Historically, most states' early attempts to control ozone focused on VOC reductions. To meet federal mandates, the commission has adopted numerous regulations controlling VOCs from marine vessel loading, vessel cleaning/degreasing, vent gases, surface coating and degreasing, and printing, among many other types of sources. These regulations have been successful in considerably decreasing VOCs and ambient ozone levels, but have not achieved compliance with the ozone standard

(*Continued*)

Texas Natural Resource Conservation Commission (Continued)

	The current ozone control strategy for Houston/Galveston (HGA), Beaumont/Port Arthur, and Dallas/Fort Worth (DFW) centers on NO_x controls. These areas are already beginning to see lowered NO_x levels as the result of the commission's rules, but the state's newly adopted regulations for stationary point sources will achieve NO_x reductions of 88–90%. In order to meet the ozone challenge in Texas' large urban areas, the commission has expanded vehicle emissions testing, regulated airport ground service equipment, and imposed seasonal operating restrictions on commercial lawn and gardening equipment and diesel construction equipment, to give a few examples. For the future, the commission is exploring innovative approaches such as emulsion fuels and catalyst after-treatment systems for diesel engines In January 1997, the Commission proposed a program that, for the first time in Texas' air pollution control history, extended beyond the confines of the urbanized areas. The purpose of the regional strategy was to reduce ozone causing compounds in the eastern half of the state in order to help reduce background levels of ozone in both nonattainment areas as well as those areas close to noncompliance for the new 8-hour ozone standard. The commission recently adopted NO_x regulations for power plants, cement kilns, and other major NO_x sources in East and Central Texas to address regional ozone. This control strategy was introduced at the commission's initiative, not in response to any specific federal requirements Meeting the 1-hour ozone standard in the state's current nonattainment areas and anticipating the 8-hour ozone standard in these areas for the state represents one of the most difficult challenges ahead for the state. Future attainment relies not only on the development of local and state control measures but also on the future federal rules involving new technologies as well. These especially involve cleaner fuels and cleaner engines for both on-road and off-road mobile sources. Until cleaner engines are available and significant fleet turnover has occurred, new and innovative solutions must be sought
Collaborative Participation in Planning and Policy Development	A number of organizations and interest groups work with the agency in developing rules and policy. In the Report to the Sunset Commission, Table 8 provides a list of subcommittees and advisory committees, and Table 24 provides a list of TNRCC Contacts. In addition, all rulemaking seeks public comment and inputs through publication in the Texas Register

(*Continued*)

Texas Natural Resource Conservation Commission (Continued)

Effectiveness of Collaborative Processes	The agency provides a number of programs to support business and regulated industry as well as the general public. For example, the Small Business and Environmental Assistance Division provides assistance to small businesses and local governments, on-site technical assistance, and environmental public awareness programs. Services include regulatory assistance seminars, technical workshops, trade fairs, waste collection events, toll-free hotline assistance, and recognition of environmental excellence. Several examples are included for your information

Utah Department of Environmental Quality

State Implementation Plan History	We have preserved the iterations of the state statute and the state implementation plans over the years, but no one has attempted to pull together information as to how and why changes were made. Note that the Utah Air Conservation Act was first adopted by the State Legislature in 1968, and air quality regulation by the State of Utah preceded the federal Clean Air Act.
Key state agencies responsible for air quality management plans	Statutory authority resides with the Air Quality Board, whose executive secretary is the director of the Division of Air Quality. Other organizations playing key roles in State Implementation Plan development are the Wasatch Front Regional Council and the Mountainland Association of Governments. Both are Metropolitan Planning Organizations with responsibilities for transportation planning. The Utah Department of Transportation also is a key player Private organizations also play a role
Funding for Air Quality Management Programs	In the early years, implementation was split about evenly among state appropriations, federal grants, and fees. Since implementation of the Operating Permits Program adopted in the CAA amendments of 1990, we have shifted to approximately 40% fees and 30% each from state appropriations and federal grants Present budget is about $8 million. We have no reliable historical information
Standards developed to evaluate SIP development	n/a

(Continued)

Utah Department of Environmental Quality (Continued)

State administrative processes for new source permitting, BACT, and revision of the AQM plans, and air quality modeling of both present and future air quality	n/a
Collaborative Participation in Planning and Policy Development	Utah Manufacturing Association, the Utah Mining Association, the Utah Petroleum Association, the Association of General Contractors, the Sierra Club, the Wasatch Clean Air Coalition, Families Against Incinerator Risk, Citizens Against Chlorine Contamination
Effectiveness of Collaborative Processes	Generally, our approach is collaborative as far as possible, in keeping with Governor Mike Leavitt's Enlibra principles (see http://www.westgov.org/wga/policy/99/99013.htm). The second principle is: Collaboration, Not Polarization—Use Collaborative Processes to Break Down Barriers and Find Solutions. For instance, a source requesting a permit for a new facility submits the necessary information to us. If DAQ finds that the facility is likely to cause exceedances of the NAAQS or does not meet other provisions of the rules, the source will be encouraged to modify the plans (change the site, add more controls, etc.) and resubmit until acceptable application is received Another example is participation in the Western Regional Air Partnership (WRAP). WRAP was created by 12 western states and an equal number of tribes to address regional air quality issues, especially focusing on implementation of the federal rule for regional haze issued on July 1, 1999. All of the WRAP's working committees and forums include members from states, tribes, federal agencies, universities, industry, environmental groups, and other interested parties

Virginia Department of Environmental Quality

State Implementation Plan History	The documentation that outlines how Virginia implemented the requirements of the Clean Air Act of 1970 is listed in 40 CFR 52.2465©. A significant change in the overall organization responsible for the implementation of the Clean Air Act in Virginia occurred in 1993, when the Department of Air Pollution Control ceased its existence as an autonomous state government agency and was subsumed as a division into the newly created Department of Environmental Quality.

(Continued)

Virginia Department of Environmental Quality (Continued)

Key state agencies responsible for air quality management plans	The key state organizations responsible for the air quality management plans in Virginia are the Department of Environmental Quality and the State Air Pollution Control Board
Funding for Air Quality Management Programs	The process was initially financed by general funds and federal grants under Section 105 of the Clean Air Act. Later, funding was added through Title V permit program fees and vehicle inspection and maintenance program fees in Northern Virginia
State's annual budget for air quality efforts required by the CAA? Historical account of expended revenue?	Virginia's air quality budget for fiscal year 2001–2002 was $14,198,314. A historical account of the funding effort is available but would require time to compile
Standards developed to evaluate SIP development	Virginia follows EPA guidance in evaluating new technologies for SIP development
State administrative processes for new source permitting, BACT, and revision of the AQM plans, and air quality modeling of both present and future air quality	Virginia follows EPA guidance and regulations in determining administrative processes for new source permitting implementation of best available control technology and revision of Virginia's air quality management program
Collaborative Participation in Planning and Policy Development	The major industrial, environmental, and professional organizations that assist Virginia in its air quality efforts include the Virginia Manufacturers Association, the American Lung Association of Virginia, the Sierra Club, BEST Consulting, Reynolds Metals, Allied Signal, Celanese Acetate, McGuire Woods Battle & Boothe, Williams Mullins Clark & Dobbins, Dominion Virginia Power, Honeywell, Ogden Martin, Columbia Gas Transmission, Old Dominion Electric Cooperative, American Electric Power Company, Potomac Electric Power, and the Southern Environmental Law Center
Effectiveness of Collaborative Processes	Virginia participates in several collaborative programs with regulated sources. For example, regulation amendments are generally developed with the assistance of *ad hoc* groups that include representatives from the regulated community

Bibliography

3M Company, *2008 Sustainability Progress* (2008).

3M Company, *2009 Environmental Solutions Catalog* (2009).

3M Company, *3M Company 1977 Annual Report* (1978).

3M Company, *3M Company 1971 Annual Report* (1972).

3M Company, 3M Company 1973 Annual Report (1974).

3M Company, *3M Company 1983 Annual Report* (1984).

3M Company, *3M Company, 1976 Annual Report* (1977).

Adelman, David E. and Kirsten H. Engel, Reorienting State Climate Change Policies to Induce Technological Change, 50 *Arizona Law Review* 835 (2008).

Adler, Jonathan H., "When is Two a Crowd? The Impact of Federal Action on State Environmental Regulation." 31 *Harvard Environmental Law Review* 1, Washington, D.C. (2007).

Aragon-Correa, Juan Alberto and Enrique A. Rubio-Lopez, "Proactive Corporate Environmental Strategies: Myths and Misunderstandings," *Long Range Planning* 40 (2007) 357–381.

Arnett, Jr., Jerome C., The EPA's Fine Particulate Matter (PM2.5) Standards, Lung Disease, and Mortality: A Failure of Epidemiology, Competitive Enterprise Institute, Washington, D.C. (2006).

Ashford, Nicholas A., "Government and Environmental Innovation in Europe and North America." *Government & Innovation* (2004): 1–20.

Automotive Industry Action Group, The AAIG Dividend: Creating Supply Chain Value (2007).

Avi-Yonah, Reuven S. and David M. Uhlmann, "Combating Global Change: Why a Carbon Tax is a Better Response to Global Warming Than Cap and Trade," 28 *Stanford Environmental Law Journal* 3 (2009).

Bachmann, John, "Will the Circle Be Unbroken: a History of the U.S. National Ambient Air Quality Standards." *Journal of the Air & Waste Management Association* 57 (2007): 652–697.

Baker, Howard H., Senator, "Cleaning America's Air – Progress and Challenges," "Remarks by Howard H. Baker, Jr.," March 9, 2005. Edmund S. Muskie Foundation. http://www.muskiefoundation.org/baker.030905.html (accessed August 2011).

Bearden, David M., Air Quality and Emissions Trading: An Overview of Current Issues, *Congressional Research Service* 98–563, 1999.

Becker, Randy, and Vernon Henderson, "Effects of Air Quality Regulations on Polluting Industries," 108 The Journal of Political Economy 379 (2000).

Billings, Leon G., "The Muskie Legacy: Policy and Politics," April 14, 2005. Edmund S. Muskie Foundation. http://www.muskiefoundation.org/billings.lecture.041405.html (accessed August 2011).

Birch, Erich Birch, "Air Quality Regulation in the United States," *Business Law Today*, July/August 2007.

Blankinship, Steve, "How Clean Air Rules Deter Equipment Investment, Power Engineering; Apr 2008; 112, 4; *ABI/INFORM Global* pg. 46.

Bleicher, Samuel A., "Economic and Technical Feasibility in Clean Air Act Enforcement Against Stationary Sources," 89 *Harvard Law Review* 316 (1975).

Blodgett, John E., Larry B. Parker, and James E. McCarthy, Air Quality Standards: The Decisionmaking Process, *Congressional Research Service*, 1998.

Blomquist, Robert F., "Government's Role Regarding Industrial Pollution Prevention in the United States." 29 *Georgia Law Review* 349 (1995).

Blomquist, Robert F., "Senator Edmund S. Muskie and the Dawn of Modern American Environmental Law: First Term, 1959–1964." 26 *William and Mary Environmental Law and Policy Review* 509 (2002).

Boczar, Barbara A., "Avenues for Direct Participation of Transnational Corporations in International Environmental Negotiations." 3 *New York University Environmental Law Journal* 1 (1994–1995).

Brian, Daniel, "Regulating Carbon Dioxide Under the Clean Air Act as a Hazardous Air Pollutant," 33 *Columbia Journal of Environmental Law* 369 (2008).

Burch, Erich, "Air Quality Regulation in the United States." *Business Law Today* 16:6 (2007).

Bushe, Craig M., "State Implementation Plans Under the Clean Air Act: Continued Enforceability as Federal Law After State Court Invalidation on State Grounds," 19 *Valparaiso University Law Review* 877 (1985).

Butler, Chad, "New Source Netting in Nonattainment Areas Under the Clean Air Act," 11 *Ecology Law Quarterly* 343 (1983–1984).

Buyssel, Kristel and Alain Verbeke, "Proactive Environmental Strategies: A Stakeholder Management Perspective," *Strategic Management Journal*. Volume 24. Pp. 453–470 (2003).

Buzbee, William W., "Clean Air Act Dynamism and Disappointments: Lessons for Climate Legislation To Prompt Innovation and Discourage Inertia. 32 *Washington University Journal of Law and Policy* 33 (2010).

Caldart, Charles C., and Nicholas A. Ashford., "Negotiation as a Means of Developing and Implementing Environmental and Occupational Health and Safety Policy." 23 *Harvard Environmental Law Review* 143 (1999).

Carlson, Ann E., Iterative Federalism and Climate Change, 103 *Northwestern University Law Review* 1097 (2009).

Chay, Kenneth, Carlos Dobkin, and Michael Greenstone, "The Clean Air Act of 1970 and Adult Mortality," *The Journal of Risk and Uncertainty*, 27:3; 279–300, 2003.

Clinton, Bill, President and Vice President Al Gore, Reinventing Environmental Regulation (March 16, 1995).

Coglianese, Cary, and Gary E. Marchant, Shifting Sands: the Limits of Science in Setting Risk Standards. *Joint Center: AEI-Brookings Joint Center for Regulatory Studies* (2003): 1–89.

Cohen, Mark A., "Firm Response to Environmental Regulation and Environmental Pressures," *Managerial and Decision Economics*, Volume 18, No. 6 (September 1996), p. 407–420.

Congressional Research Service, Clean Air Act: A Summary of the Act and Its Major Requirements. note 29, at p. 10, 2005.

Congressional Research Service, Clean Air Permitting: Status of Implementation and Issues CRS-2 (2006).

Cooney, Stephen and Brent D. Yacobucci, U.S. Automotive Industry: Policy Overview and Recent History CRS-95 (Congressional Research Service April 25, 2005).

Costle, Douglas M., Oral History Interview, http://nepis.epa.gov/Exe/ZyPURL.cgi?Dockey=1000494F.txt (accessed August 2011).

Craig, Robin Kundis, "Removing 'the Cloak of a Standing Inquiry': Pollution Regulation, Public Health, and Private Risk in the Injury-in-Fact Analysis." *Cardoza Law Review* (October 2007).

Crider, Kay M., Interstate Air Pollution: Over a Decade of Ineffective Regulation, 64 *Chicago-Kent Law Review* 619 (1988).
Croci, Edoardo, Negotiated Regulation, Implementation and Compliance in the United States. Vol. 43. Netherlands: Springer, 2005.
Currie, David P., "Federal Air-Quality Standards and Their Implementation," 1976 *American Bar Foundation Journal* 365 (1976).
Currie, David P., "Relaxation of Implementation Plans under the 1977 Clean Air Act Amendments," 78 *Michigan Law Review* 155 (1979).
Davis, Devra, *When Smoke Ran Like Water* (2002 Basic Books), New York, NY.
Dockery, D.W. and C.A. Pope, III, "Acute Respiratory Effects Of Particulate Air Pollution," *Annual Review of Public Health*. 1994. 15:107–32.
Doremus, Holly and W. Michael Hanemann, Of Babies and Bathwater: Why the Clean Air Acts Cooperative Federalism Framework is Useful for Addressing Global Warming, 50 *Arizona Law Review* 799 (2008).
Dow Chemical Company, *2009 Annual Report* (2010).
Dow Chemical Company, *2010 Annual Report* (2011).
Dow Chemical Company, *2010 Global Reporting Initiative Report/Annual Sustainability Report* (2010).
Dowd, Richard M., The Role of Science in EPA Decision Making, Environmental Science and Technology, Volume 15, No. 10, October 1981. pp. 1137–1141.
E. I. DuPont & Nemours Company, *Annual Report 1970* (1971).
E. I. DuPont & Nemours Company, *DuPont Annual Report 1971* (1972).
E. I. DuPont & Nemours Company, *DuPont Annual Report 1972* (1973).
E. I. DuPont & Nemours Company, *2015 Sustainability Goals* (2006). http://www2.dupont.com/Sustainability/ en_US/assets/downloads/FINAL_BROCHURE_9.28.06.pdf
E. I. DuPont & Nemours Company, *2010 Annual Review* (2011)
E. I. DuPont & Nemours Company, *2010 Sustainability Progress Report* (2010).
E. I. DuPont & Nemours Company, *DuPont 1980 Annual Report* (1981).
E. I. DuPont & Nemours Company, *DuPont 1982 Annual Report* (1983).
E. I. DuPont & Nemours Company, *DuPont 1984 Annual Report* (1985).
E. I. DuPont & Nemours Company, *DuPont 1996 Annual Report* (1997).
E. I. DuPont & Nemours Company, *DuPont 1999 Annual Report* (2000).
E. I. DuPont & Nemours Company, *DuPont 2006 Annual Review* (2007).
E. I. DuPont & Nemours Company, *DuPont 2007 Annual Review* (2008).
E. I. DuPont & Nemours Company, *DuPont Annual Report 1971* (1972).
E. I. DuPont & Nemours Company, *DuPont Annual Report 1973* (1974).
E. I. DuPont & Nemours Company, *DuPont Annual Report 1974* (1975).
E. I. DuPont & Nemours Company, *DuPont Annual Report 1975* (1976).
E. I. DuPont & Nemours Company, *DuPont Annual Report 1976* (1977).
E. I. DuPont & Nemours Company, *DuPont Annual Report 1979* (1980).
E. I. DuPont & Nemours Company, *DuPont Annual Report 1989* (1990).
E. I. DuPont & Nemours Company, *DuPont Annual Report 1990* (1991).
E. I. DuPont & Nemours Company, *DuPont Annual Report 1991* (1992).
E. I. DuPont & Nemours Company, *DuPont: A Global Entrepreneur Annual Report 1988* (1989).
E. I. DuPont & Nemours Company, *Sustainable Growth Through Science – 2005 Annual Review* (2006).

Easton, Eric B. and Francis J. O'Donnell, The Clean Air Act Amendments of 1977: Refining the National Air Pollution Control Strategy, 27 *Journal of the Air Pollution Control Association*, 943, 943–47 (1977).

Easton, Eric B. and Francis J. O'Donnell, "The Clean Air Act Amendments of 1977: Refining the National Air Pollution Control Strategy." *Journal of Air Pollution Control* 27 (1977): 943–47.

Eaton, Seth W., Winter is Frigid, So I Say Bring on the Greenhouse Effect! A Legal and Policy and Discussion of the Strategies the United States Must Employ to Combat Global Warming, 35 *Pepperdine Law Review* 787 (2008).

Ember, Lois R., "EPA Administrators Deem Agency's First 25 Years Bumpy but Successful," *Chemical and Engineering News*, Oct. 25, 1995.

Epstein, Richard A., Carbon Dioxide: Our Newest Pollutant, 18 *Suffolk University Law Review* 797 (2010).

Executive Office of the President, President's Advisory Council on Executive Organization, "Memorandum for the President, Subject: Federal Organization for Environmental Protection," April 29, 1970.

Exxon Annual Report 1998 (1999).

Exxon Corporation 1975 Annual Report (1976).

Exxon Corporation 1976 Annual Report (1977).

Exxon Corporation 1993 Annual Report (1994).

Exxon Corporation 1994 Annual Report (1995).

Exxon Corporation 1995 Annual Report. (1996).

Exxon Mobil, *2006 Citizenship Report* (2006).

Exxon Mobil, *2007 Corporate Citizenship Report* (2007).

Exxon Mobil, *2008 Corporate Citizenship Report* (2009).

Exxon Mobil, *2009 Corporate Citizenship Report* (2009).

Exxon Mobil, *2009 Summary Annual Report* (2010).

Exxon Mobil, *2010 Corporate Citizenship Report* (2010).

Exxon Mobil, *Managing for Environmental Excellence* (Nov. 2004).

Exxon Mobil, *Operations Integrity Management System* (2009).

Exxon Mobil, *Summary Annual Report 2000* (2001).

Fiorino, Daniel J., "Toward a New System of Environmental Regulation: the Case for an Industry Sector Approach." *Environmental Law* 26 (1996): 1–3.

Ford Motor Company, *1998 Annual Report* (1999).

Ford Motor Company, *2002 Corporate Citizenship Report* (2003).

Ford Motor Company, *2007/8 Blueprint for Sustainability* (2007).

Ford Motor Company, *2007/8 Blueprint for Sustainability* (2008).

Ford Motor Company, *2008 Annual Report* (2009).

Ford Motor Company, *2009–2010 Sustainability Report* (2010).

Ford Motor Company, *Ford 1971 Annual Report* (1972).

Ford Motor Company, *Ford Annual Report 1977* (1978).

Ford Motor Company, *Connecting With Society: Our Learning Journey* (2008).

Forswall, Clayton D. and Kathryn E. Higgins, Clean Air Act Implementation in Houston: An Historical Perspective 1970–2005. Environmental and Energy Systems Institute, Shell Center for Sustainability, Rice University, February 2005.

Fox, David, "DOE Awards WSI for Outstanding Safety and Health Programs." *Department of Energy: VPP StarBurst* (2001): 1–20.

Franz, Neil, "Cleaning Up the Clean Air Act." *Chemical Engineering* 108 (2001): 1–3.

Furlong, Scott R., "Reinventing Regulatory Development at the Environmental Protection Agency," *Policy Studies Journal* Vol 23, No. 3,1995 (466–482).

Gauna, Eileen, "Major Sources of Criteria Pollutants in Nonattainment Areas: Balancing the Goals of Clean Air, Environmental Justice, and Industrial Development." *Hastings West-Northwest Journal of Environmental Law and Policy* (1996).

Geiger, Jeffrey, "Canary in a Coal Mine? Federalism and the Failure of the Clean Air Act Amendments of 1990." *William and Mary Environmental Law and Policy Review* (1995).

General Accounting Office, *Air Pollution: Status of Implementation and Issues of the Clean Air Act Amendments of 1990* (Apr 2000).

General Accounting Office, *CLEAN AIR ACT: EPA Has Completed Most of the Actions Required by the 1990 Amendments, but Many Were Completed Late* (May 2005).

General Accounting Office, *Clean Air Rulemaking: Tracking System Would Help Measure Progress of Streamlining Initiatives* 18 (March 1995).

General Accounting Office, *Environmental Protection: Collaborative EPA-State Effort Needed to Improve New Performance Partnership System* (June 1999).

General Accounting Office, *Status of Implementation and Issues of the Clean Air Act Amendments of 1990* (April 2000).

General Accounting Office, *Air Pollution: EPA's Strategy and Resources May Be Inadequate to Control Air Toxics* (1991).

General Accounting Office, *Air Pollution: EPA's Strategy and Resources May Be Inadequate to Control Air Toxics GAO/RCED-91-143* (June 1991).

General Accounting Office, *Air Pollution: EPA's Strategy and Resources May Be Inadequate to Control Air Toxics.* (June 1991).

General Accounting Office, *Air Pollution: Reduction in EPA's Air Quality Budget* (Nov 1994).

General Accounting Office, *Air Pollution: Reduction in EPA's Air Quality Budget* (U.S. G.A.O. Nov 1994).

General Accounting Office, *Air Pollution: Status of Implementation and Issues of the Clean Air Act Amendments of 1990* (April 2000).

General Accounting Office, *Challenges Facing EPA's Efforts to Reinvent Environmental Regulation* (July 1997).

General Accounting Office, *Clean Air Act: EPA Should Improve the Management of its Air Toxics Program* (June 2006).

General Accounting Office, *Clean Air Rulemaking: Tracking System Would Help Measure Progress of Streamlining Initiatives.* March 1995.

General Accounting Office, *Report to the Chairman, Subcommittee on Clean Air, Wetlands, Private Property, and Nuclear Safety, Committee on Environment and Public Works, U. S. Senate, Air Pollution: Status of Implementation and Issues of the Clean Air Act Amendments of 1990, April 2000,* GAO/RCED-00-72.

General Accounting Office, *Report to the Chairman. Subcommittee on Oversight and Investigations, Committee on Energy and Commerce, House of Representatives. Air Pollution: EPA's Strategy and Resources May Be Inadequate to Control Air Toxics.* GAO/RCED-91-143, June 1991: 3.

General Accounting Office, *CLEAN AIR ACT: EPA Should Improve the Management of Its Air Toxics Program,* at p. 5 (June 2006).

Gerard, David and Lester B. Lave, "Implementing Technology-Forcing Policies: The 1970 Clean Air Act Amendments and the Introduction of Advanced Automotive Emissions Controls in the United States," Technology Forecasting & Social Change 72 (2005) p. 761–778.

Giovinazzo, Christopher, "Defending Overstatement: The Symbolic Clean Air Act and Carbon Dioxide," 30 *Harvard Environmental Law Review* 99 (2006).

Greenstone, Michael, "The Impacts of Environmental Regulations on Industrial Activity: Evidence from the 1970 and 1977 Clean Air Act Amendments and the Census of Manufacturers," 110 *Journal of Political Economy* 1175 (2002).

Greenstone, Michael, "Estimating Regulation-Induced Substitution: the Effect of the Clean Air Act on Water and Ground Pollution." 93 *The American Economic Review* 1 (2003).

Greenstone, Michael, "The Impacts Of Environmental Regulations On Industrial Activity: Evidence From The 1970 And 1977 Clean Air Act Amendments And The Census Of Manufactures," 110 *Journal of Political Economy* 1175 (2002).

Grumet, Jason S., "Old West Justice: Federalism and Clean Air Regulation 1970–1998." *Tulane Environmental Law Journal* (Summer 1998).

Hahn, Robert W., "Regulatory Reform: Assessing the Government's Numbers." *Joint Center: AEI-Brookings Joint Center for Regulatory Studies* 99 (1999): 1–54.

Heaton, George R., and R D. Banks, "Toward a New Generation of Environmental Technology: the Need for Legislative Reform." Journal of Industrial Ecology 1 (1997): 23–32. *MIT Press Journals.*

Heinzerling, Lisa, "Ten Years After the Clean Air Act Amendments: Have We Cleared the Air?" 20 *St. Louis University Public Law Review* 121 (2001).

Heinzerling, Lisa, "Federal, State, and Litigation Initiatives: Climate Change and the Clean Air Act," 42 *University of San Francisco Law Review* 111 (2007).

Helfand, William H. Jan Lazarus, and Paul Theerman, "Donora, Pennsylvania: An Environmental Disaster of the 20th Century," 91 *American Journal of Public Health* 553 (April 2001).

Heller, Karen and Ronald Begley, "Redefining the Role and Obligations of an Industry," *Chemistry Week*, July 6-July 13, 1994.

Henderson, James A. and Richard N. Pearson, "Implementing Federal Environmental Policies: The Limits of Aspirational Commands," 78 *Columbia Law Review* 1429 (1978).

Ho, Mun S. and Dale W. Jorgenson, "Market-based policies for air-pollution control," *Harvard Magazine*, September-October 2008, pp. 32.

Holtkamp, James A., The Clean Air Act, Holland & Hart LLP, 1 Aug 2003.

ICF Resources Inc. and Smith Barney, Harris Upham, and Company, Business Opportunities of the New Clean Air Act: The Impact of the CAAA of 1990 on the Air Pollution Control Industry, August 2002.

Imsland, Lori C., "How Much Would You Pay For Clean Air? The Role of Costs/Benefit Analysis in Setting NAAQS," 9 *Missouri Environmental Law and Policy Review* 44 (2002).

Irene Henriques, Irene and Perry Sadorsky, "The Relationship between Environmental Commitment and Managerial Perceptions of Stakeholder Importance," *The Academy of Management Journal*, Vol. 42, No. 1. (Feb., 1999), pp. 87–99.

Jeffords, James M., and Frank R. Lautenberg, Chemical Regulation: Options Exist to Improve EPA's Ability to Assess. *United States Government Accountability Office (GAO).* Washington, D.C., 2005.

Johnston, Jason S., "Climate Change Confusion and the Supreme Court: the Misguided Regulation of Greenhouse Gas Emissions under the Clean Air Act" *Scholarship at Penn Law Paper* 209: 2008).

Johnston, Jason Scott, "Climate Change Confusion and the Supreme Court: The Misguided Regulation of Greenhouse Gas Emissions Under the Clean Air Act," 84 *Notre Dame Law Review* 1 (2008).

Jolish, Taly L., "Negotiating the Smog Away," 18 *Virginia Environmental Law Journal* 305 (1999).

Kahn, Shulamit, and Christopher R. Knittel, "The Impact of the Clean Air Act Amendments of 1990 on Electric Utilities and Coal Mines: Evidence From the Stock *Market." University of California Energy Institute: Center for the Study of Energy Markets* (2003): 1–30.

Kennedy, Harold W. and Martin E. Weekes, "Control of Automobile Emissions—California's Experience and the Federal Legislation," 33 *Law and Contemporary Problems* 297 (1968).

Korostash, Yekaterina, "EPA's New Regulatory Policy: Two Steps Back." 5 *North Carolina Journal of Law & Technology* 295 (2004).

Kraft, Michael E. and Denise Scheberle, "Environmental Federalism at Decade's End: New Approaches and Strategies," 28 *Publius: The Journal of Federalism* 1 (Winter 1998).

Krupnick, Alan, and Richard Morgenstern, "The Future of Benefit – Cost Analyses of the Clean Air Act." *Annual Review of Public Health* 23 (2002): 427–448.

Lave, Lester B., "EPA's Proposed Air Quality Standards: Clean Air Sense." *Environment, Energy Security, Regulation, Business, Environmental Regulation* (1997).

Lee, Amanda L., "The Clean Air Act Amendments and Firm Investment in Pollution Abatement Equipment," 80 *Land Economics* 433 (2004).

Lewis, Jack, "The Birth of EPA," *EPA Journal* (November 1985).

Lieberman, Ben, Kay Jones, and Indur Goklany, The Clean Air Act. *Competitive Enterprise Institute.* Washington, D.C., 1999.

Lippman, M. and R.B. Schlesinger, "Toxicological Bases for the Setting of Health-Related Air Pollution Standards" *Annual Review of Public Health* 2000. 21:309–33.

List, John A., Daniel L. Millimet, and Warren McHone, "The Unintended Disincentive in the Clean Air Act," *Advances in Economic Analysis & Policy* Volume 4, Issue 2 (2004) pp. 1–26.

Logan, Andrew and David Grossman, ExxonMobil's Corporate Governance on Climate Change (*CERES* 2006).

Lutter, Randall W., "An Analysis of the Use of EPA's Benefit Estimates in OMB's Draft Report on the Costs and Benefits of Regulation." *Joint Center: AEI-Brookings Joint Center for Regulatory Studies* 98 (1998): 1–18.

Lyon, Thomas P. and John W. Maxwell, Corporate Environmentalism and Public Policy, *Cambridge University Press*, 2004.

Martin, Robert and Lloyd Symington, "A Guide to the Air Quality Act of 1967" 33 *Law and Contemporary Problems* 241 (1968).

Martineau Jr., Robert J. and David P. Novello, The Clean Air Act Handbook, Second Edition. *American Bar Association* 2004.

McCarthy, James E., Clean Air Act: A Summary of the Act and Its Major Requirements (*Congressional Research Service*: May 2005).

McCarthy, James E., Highway Fund Sanctions for Clean Air Act Violations, *Congressional Research Service*, 1997.

McCrory, Martin A. and Eric L. Richards, "Clearing the Air: The Clean Air Act, GATT And The WTO's Reformulated Gasoline Decision." 17 *UCLA Journal of Environmental Law and Policy* 1 (1998–99).

McCubbin, Patricia Ross, "Proposed Endangerment and Cause or Contribute Findings for Greenhouse Gases Under Section 202(a) of the Clean Air Act," 33 *Southern Illinois University Law Journal* 437, 438–439 (2009).

McCubbin, Patricia, "The Risk in Technology-Based Standards," 16 *Duke Environmental Law & Policy Forum* 1 (2005).

McGarity, Thomas O., "The Clean Air Act at a Crossroads: Statutory Interpretation and Longstanding Administrative Practice in the Shadow of the Delegation Doctrine." 9 *N.Y.U. Environmental Law Journal* 1 (2000).

McKinstry, Jr., Robert B., "Laboratories for Local Solutions for Global Problems: State, Local, and Private Leadership in Developing Strategies to Mitigate the Causes and Effects of Climate Change," 12 *Penn State Environmental Law Review*, 15 (2004).

McNollgast, "Legislative Intent: the Use of Positive Political Theory in Statutory Interpretation." *Law and Contemporary Problems* 57 (1994): 3–37.

Melnick, R. Shep, Regulation and the Courts: The Case of the Clean Air Act. Washington, D.C.: *Brookings Institute*, 1983.

Menz, Fredric C., Mobile Source Pollution Control in the United States and China. *Center for International Climate and Environmental Research*, Oslo, Norway, April 2002.

Merrifield, John, "A Critical Overview of the Evolutionary Approach to Air Pollution Abatement Policy," 9 *Journal of Policy Analysis and Management* 367 (1990).

Molina, Louis T. and Mario J. Molina, Air Quality in the Mexico Megacity: 31 (2002).

Muskie, Edmond S., Senator, "NEPA to CERCLA, The Clean Air Act: A Commitment to Public Health" *The Environmental Forum* (January/February 1990).

National Academy of Sciences, Committee on Air Quality Management in the United States, Air Quality Management in the United States (*National Academy of Sciences*, 2004), Washington D.C.

National Petrochemical and Refiners Association, *Annual Report 2008* (2008).

National Research Council, Environmental Studies Board, On Prevention of Significant Deterioration of Air Quality. *National Academy Press*, Washington D.C. (1981).

Northeast States for Coordinated Air Use Management, NESCAUM, 1967-2007, Forty Years. (2007).

Northeast States for Coordinated Air Use Management., Public Health Benefits of Reducing Ground-level Ozone and Fine Particulate Matter in the Northeast U.S. (2008).

Norvick, Steve and Bill Westerfield, "Whose SIP Is It Anyway? State-Federal Conflict in Clean Air Act Enforcement." 18 *William and Mary Journal of Environmental Law* 245 (1994).

O'Brien, David M., "Regulation and the Courts: the Case of the Clean Air Act." 78 *The American Political Science Review* 804 (1984).

Oren, Craig N., "Prevention of Significant Deterioration: Control-Compelling Versus Site-Shifting," 74 *Iowa Law Review* 1 (October 1988).

Patton, Vickie, Curbing Interstate Air Pollution, *Forum for Applied Research and Public Policy*, Fall 2001, pp. 21–27.

Pedersen, William F., Jr., "Why the Clean Air Act Works Badly," 129 *University of Pennsylvania Law Review* 1059 (1981).

Peterson, Thomas D., Robert B. McKinstry, Jr. and John C. Dernbach, "Global Climate Change: Individual, Private Sector, and State Responses: Developing a Comprehensive Approach to Climate Change Policy in the United States that Fully Integrates Levels of Government and Economic Sectors," 26 *Virginia Environmental Law Journal* 227 (2008).

Peterson, Thomas D., Robert B. McKinstry, Jr. and John C. Dernbach, "Federal Climate Change Legislation As if the States Mattered," *Natural Resources and Environment*, Winter 2008, pp. 3–8.

Portney, Paul R., "Policy Watch: Economics and the Clean Air Act," 4 *Journal of Economic Perspectives* 173 (1990).

Poulton, Michael T., "Particles Of What? A Call For Specificity in Airborne Particulate Regulation," 51 *Jurimetrics* 61 (2010).

Powell, Mark R., Science at EPA: Information in the Regulatory Process. Washington, D.C.: Resources for the Future, 1999.

Procter & Gamble, *2010 Sustainability Report* (2010).

Procter & Gamble, *The Procter and Gamble Annual Report For the Year Ended June 30, 1970* (1970).

Procter & Gamble, *Procter & Gamble, Global Opportunities Global Growth, 1995 Annual Report* (1995).

Procter & Gamble, *Procter and Gamble Annual Report for the Year ending Jun 30, 1972* (1972).

Procter & Gamble, *1999 Sustainability Report* (1999).

Procter & Gamble, *2000 Sustainability Report* (2001).

Raiders, Rich, "How EPA Could Implement a Green House Gas NAAQS", 22 *Fordham Environmental Law Review* 233 (2011).

Randle, Russell V., "Forcing Technology: The Clean Air Act Experience." 88 *Yale Law Journal* 1713 (1979).

Reilly, William K., "The New Clean Air Act: An Environmental Milestone," *EPA Journal*, January-February 1991.

Reitze, Arnold W., "The Role of the 'Region' in Air Pollution Control," 20 *Case Western Reserve Law Review* 809 (1969).

Reitze, Arnold W., "Air Quality Protection Using State Implementation Plans – Thirty-Seven Years Of Increasing Complexity." 15 *Villanova Environmental Law Journal* 209 (2004).

Reitze, Arnold W. Jr., "A Century Of Air Pollution Control Law: What's Worked; What's Failed; What Might Work, 21 *Environmental Law* 1549 (1995).

Reitze, Arnold W., Jr., "Federal Control of Carbon Dioxide: What are the Options?" 36 *Boston College Environmental Affairs Law Review* 1 (2009).

Reitze, Arnold, Air Pollution Control Law: Compliance and Enforcement, *Environmental Law Institute*, Washington, D.C. 2001.

Ringquist, Evan J., Environmental Protection at the State Level (*M.E. Sharpe*, 1993).

Robinson, James C. & William S. Pease, "From Health-Based to Technology-Based Standards for Hazardous Air Pollutants, 81 *American Journal of Public Health* 1518 (1991).

Rogers, Paul, "The Clean Air Act of 1970," *EPA Journal* (Jan/Feb 1990).

Rohm & Haas, *Rohm and Haas 1976 Annual Report* (1977).

Rohm & Haas, *Rohm and Haas 1979 Annual Report* (1980).

Rohm & Haas, *Rohm and Haas 1985 Annual Report* (1986).

Rohm & Haas, *Rohm and Haas 2006 EHS and Sustainability Report* (2007).

Rohm & Haas, *Rohm and Haas Annual Report 1970* (1971).

Rohm & Haas, *Rohm and Haas Annual Report 1971* (1972).

Rohm & Haas, *Rohm and Haas Annual Report 1981* (1982).

Rohm & Haas, *Rohm and Haas Annual Report 1992* (1993).

Rohm & Haas, *Rohm and Haas Annual Report 1995* (1996).

Rohm & Haas, *Rohm and Haas Annual Report 1997* (1998).

Rohm & Haas, *Rohm and Haas Annual Report 1998* 21 (1999).

Rohm & Haas, *Rohm and Haas Annual Report 2000* 23 (2001).

Rohm & Haas, *Rohm and Haas Annual Report 2003* 12 (2004).

Rohm & Haas, *Rohm and Haas Company 1974 Annual Report* (1975).

Rohm & Haas, *Rohm and Haas Company Annual Report 1972* (1973).

Rohm & Haas, *Rohm and Haas Company Annual Report 1973* (1974).

Rosenberg, Ronald H., "Cooperative Failure: An Analysis of Intergovernmental Relationships and the Problem of Air Quality Non-Attainment." *1990 Annual Survey of American Law.* 13 (1991).

Ross, Heather L., "The Search for an Intelligible Principle: Setting Air Quality Standards Under the Clean Air Act." *Resources for the Future* (2000).

Saha, Bansari, Barry Galef, Lou Browning and Jim Staudt, "The Clean Air Act Amendments: Spurring Innovation and Growth While Cleaning the Air." *ICF Consulting* (2005), pp. 1–43.

Schoenbrod, David, "Goals Statutes or Rules Statutes: The Case of the Clean Air Act," 30 *UCLA Law Review* 740 (1983).

Schoenbrod, David, "The EPA's Faustian Bargain." *Regulation* pp. 36–42 (Fall 2006).

Schwartz, Eric, "Carbon Dioxide and the Clean Air Act," 4 *Cardozo Public Law, Policy & Ethics Journal* 779 (2006).

Scott, Janea, John Balbus, Jana Milford, Vickie Patton, Nancy Spencer and Rachel Zwillinger, The Clean Air Act At 35: Preventing Death And Disease From Particulate Pollution. *Environmental Defense Fund*, Washington, D.C. 2005.

Scott, John T., "Environmental Research Joint Ventures Among Manufacturers," 11 *Review of Industrial Organization* 655 (1996).

Siegler, Ellen, "Regulatory Negotiations and Other Rulemaking Processes: Strengths and Weaknesses from an Industry Viewpoint," 46 *Duke Law Journal* 1429 (1997).

Solomon, S., Qin, D., Manning, M., Chen, Z., Marquis, M., Averyt, K.B., Tignor M. and Miller, H.L. Climate Change 2007: The Physical Science Basis. Contribution of Working Group I to the Fourth Assessment Report of the Intergovernmental Panel on Climate Change *Cambridge University Press*, Cambridge, UK.

Standard Oil Company (New Jersey), *1970 Annual Report* (1971).

Stewart, Richard B., "Pyramids of Sacrifice? Problems of Federalism in Mandating State Implementation of National Environmental Policy." 68 *Yale Law Journal* 1196 (1977).

Stone, Brian Jr., Adam C. Mednick, Tracey Holloway, and Scott N. Spak, "Is Compact Growth Good for Air Quality?" 73 *Journal of the American Planning Association* 404 (2007).

Stone, Jr., Brian, "Air Quality by Design." 23 *Journal of Planning Education and Research* 17 (2003).

Strasser, Kurt A., "Cleaner Technology, Pollution Prevention and Environmental Regulation." 9 *Fordham Environmental Law Journal* 21 (1997).

Sunstein, Cass R., "Is the Clean Air Act Unconstitutional?" 98 *Michigan Law Review* 303 (1999).

Texas Instruments, *1991 Annual Report* (1992).

Texas Instruments, *1994 Annual Report* (1995).

Texas Instruments, *1990 Annual Report* (1991).

Texas Instruments, *2008 Corp Citizenship Report Summary* (2008).

Texas Instruments, *2009 Corp Citizenship Report Summary* (2009).

Texas Instruments, *Building a Better Future, Environmental, Safety, and Health 2005 Review* (2006).

Texas Instruments, *Environmental, Safety, and Health – 2004 Review* (2005).

Texas Instruments, *Environmental, Safety, and Health – 2004 Review*, at 6.

The Boeing Company, *2007 Annual Report* (2008).

The Boeing Company, *1995 Annual Report* (1996).

The Boeing Company, *1998 Annual Report* (1999).

The Boeing Company, *Boeing 2008 Environment Report* (2008).

The Boeing Company, *Boeing 2008 Environment Report* (2008).

The People's Republic of China, *National Implementation Plan for the Stockholm Convention on Persistent Organic Pollutants* (April 2007).

The State Environmental Protection Administration of China & The World Bank Rural Development, Natural Resources and Environment Management Unit, East Asia and Pacific Region, *Cost of Pollution in China: Economic Estimates of Physical Damage* (2007).

The White House Council on Environmental Quality, Progress Report of the Interagency Climate Change Adaptation Task Force, *Recommended Actions in Support of a National Climate Change Adaptation Strategy*, October 5, 2010.

Train, Russell E., "EPA's Task," *EPA Journal*, Nov./Dec. 1980.

U.S. Environmental Protection Agency, Office of Inspector General, Audit Report, The Effectiveness and Efficiency of EPA's Air Program. Report No. E1KAE4-05-0246-8100057 (1998): 3.

U.S. EPA Air Quality Management Work Group, *Recommendations to the Clean Air Act Advisory Committee* (2005).

U.S. EPA Clean Air Act Advisory Committee, *The Clean Air Act of 1990: An Introductory Guide to Smart Implementation* (1992).

U.S. EPA Clean Air Act Implementation Task Force, *Report to the Deputy Administrator* (EPA 1990).

U.S. EPA, *Fiscal Year 2010 Agency Financial Report* at p. 3 (2010).

U.S. EPA New England, *State of the New England Environment*, 1970–2000 (2008).

U.S. EPA Office of Air and Radiation, *Implementation Strategy for the Clean Air Act Amendments of 1990* (Update 1992).

U.S. EPA Office of Air and Radiation, *Report of the Office of Air and Radiation to Administrator William K. Reilly, Implementing the 1990 Clean Act: The First Two Years* (EPA 1992).

U.S. EPA Office of Compliance Assistance, *Compliance Assistance Activity Plan Fiscal Year 2001* (2001).

U.S. EPA Office of Compliance Sector Notebook Project, *Profile of the Dry Cleaning Industry* (EPA September 1995).

U.S. EPA Office of Inspector General, Audit Report, *The Effectiveness and Efficiency of EPA's Air Program* (1998).

U.S. EPA Office of Inspector General, *Audit Report, The Effectiveness and Efficiency of EPA's Air Program Report No. E1KAE4-05-0246-8100057* (1998).

U.S. EPA Office of Policy, Economics, and Innovation, Stakeholder Involvement & Public Participation at the U.S. EPA, *Lessons Learned, Barriers, & Innovative Approaches* (January 2001).

U.S. EPA Region 5, *Remember the Past, Protect the Future, Chicago, Illinois, Region 5 EPA* (2000).

U.S. EPA Science Advisory Board, *FY 2001 Annual Staff Report: Expanding Expertise and Experience* (December 2001).

U.S. EPA, "History: Ash Council Memo, Subject: Federal Organization for Environmental Protection," *Executive Office of the President, President's Advisory Council on Executive Organization, Memorandum for the President*. April 29, 1970.

U.S. EPA, "Region Functional Statements," *Organization and Functions Manual*, November 2000.

U.S. EPA, 1974 Legal Supplement, Supplement II, Volume 1 (Air) Jan. 1974).

U.S. EPA, *2006–2011 EPA Strategic Plan Charting Our Course*. 14 (Oct 2006).

U.S. EPA, *Alternative Control Technology Documents for Bakery Oven Emissions* (1992).

U.S. EPA, Clean Air Act, *Key Stakeholders' Views on Revisions to the New Source Review Program* (February 2004).

U.S. EPA, *Engaging the American People, A Review of EPA's Public Participation Policy and Regulations with Recommendations for Action* at p. 20 (2000).

U.S. EPA, *Environmental Protection Agency: A Progress Report, December 1970-June 1972*, pp. 1–2 (Washington, D.C. 1972).

U.S. EPA, *EPA Progress Report December 1970 to June 1972* (1972).

U.S. EPA, *Fiscal Year 2010, Agency Financial Report, 2011* (2011).

U.S. EPA, *FY 2006 Annual Plan* (2006).

U.S. EPA, *FY EPA 2010 Budget in Brief*, (2009).

U.S. EPA, *Guidelines for Air Quality Maintenance Planning and Analysis, Volume 2: Plan Preparation* (1974).

U.S. EPA, *Innovating for Environmental Results: A Strategy to Guide the Next Generation at EPA* (April 2002).

U.S. EPA, *Legal Compilation – Supplement II* (1974).

U.S. EPA, *Legal Compilation* 1973.

U.S. EPA, *Office of Pollution Prevention and Toxics, Pollution Prevention Incentives for States*, Spring 1994.

U.S. EPA, *Office of the Chief Financial Officer, FY 2008 in Brief, February 2007*. Report Number EPA-205-S-07-001.

U.S. EPA, *Prevention of Significant Deterioration Workshop Manual* (October 1980).

U.S. EPA, *Progress in the Prevention and Control of Air Pollution in 1980 and 1981, Annual Report of the Administrator of the Environmental Protection Agency to the Congress of the United States* (EPA 1981).

U.S. EPA, *Progress in the Prevention and Control of Air Pollution in 1982, Annual Report of the Administrator of the Environmental Protection Agency to the Congress of the United States* (EPA 1982).

U.S. EPA, *Progress in the Prevention and Control of Air Pollution in 1983* (EPA 1985).

U.S. EPA, *Progress in the Prevention and Control of Air Pollution in 1985* (EPA 1987).

U.S. EPA, *Progress in the Prevention and Control of Air Pollution in 1987* (EPA 1989).

U.S. EPA, *Progress in the Prevention and Control of Air Pollution in 1988* (EPA 1990).

U.S. EPA, *Progress in the Prevention and Control of Air Pollution in 1986* (EPA 1988).

U.S. EPA, *Progress Report, 1970–1972*.

U.S. EPA, *Scientific Seminar on Automotive Pollutants, February 10–12, 1975*, Washington, D.C.

U.S. EPA, *The Benefits and Costs of the Clean Air Act 1970–1990 EPA Report to Congress*, October 1997.

U.S. EPA, *A Progress Report December 1970 to June 1972* (EPA 1972).

U.S. EPA, *The Clean Air Act of 1990, a Primer on Consensus Building*. Washington: GPO, 1990.

U.S. EPA, *Title V Task Force, Final Report to the Clean Air Act Advisory Committee: Title V Implementation Experience* (April 2006).

U.S. Global Change Research Program, *Global Climate Change Impacts in the United States*, Cambridge University Press, New York 2009.

United Stated House of Representatives, Committee on Oversight and Government Reform, *Political Interference with Climate Change Science Under the Bush Administration*, December 2007.

United States Council for Automotive Research, *Power of Automotive Collaboration* (2010).

United States Department of State, *U.S. Climate Action Report, 2010*. Washington: Global Publishing Services, June 2010.

United States Environmental Protection Agency, *The Plain English Guide to the Clean Air Act*, 2007.

United States Environmental Protection Agency, "Region Functional Statements," (Sept. 2007).

US EPA Science Policy Council. Science Policy Council Handbook. 2000.

Vandenberg, John J., "The Role of Air Quality Management Programs in Improving Public Health: A Brief Synopsis," *Journal of Allergy and Clinical Immunology*, Vol. 15, Number 2, pp. 334–336.

Vietor, Richard H. K., "The Evolution of Public Environmental Policy: The Case of 'Non-Significant Deterioration," *Environmental Review* 3 (Winter 1979).

Wagner, Wendy, "The "Bad Science" Fiction: Reclaiming the Debate over the Role of Science in Public Health and Environmental Regulation," *Law and Contemporary Problems*, Vol. 66, No. 4, Science in the Regulatory Process (Autumn, 2003), pp. 63–133.

Weber, Edward P. and Anne M. Khademian, "From Agitation to Collaboration: Clearing the Air through Negotiation." 57 *Public Administration Review* 396 (1997).

Weiner, Jonathan B., "Think Globally, Act Globally, the Limits of Local Climate Policies," 155 *University of Pennsylvania Law Review* 1961 (2007).

Wetering, Sarah B. and Matthew McKinney, "The Role of Mandatory Dispute Resolution In Federal Environmental Law: Lessons From The Clean Air Act," 21 *Journal of Environmental Law and Litigation* 1 (2006).

Williams, Dennis C., "The Guardian: EPA's Formative Years, 1970–1973." September 1993. *EPA 202-K-93-002.*

Wise Sullivan, Johanna L., "The Limited Power of States to Regulate Nonroad Mobile Sources Under the Clean Air Act." 34 *Boston College Environmental Affairs Law Review* 1 (2007).

Wisman, Phil, "EPA History (1970–1985)," *EPA Journal* (November 1985).

Witham, Lyle, A Summary of the Development of the Clean Air Act 3–11 (*North Dakota Office of the Attorney General* Aug 2005).

Wolman, M. Gordon, On Prevention of Significant Deterioration of Air Quality, by Environmental Studies Board. 1 *Journal of Policy Analysis and Management* 2 (2002).

Wood, B. Dan, "Modeling Federal Implementation as a System: The Clean Air Case," *American Journal of Political Science*, Vol. 36, No. 1. (Feb., 1992), pp. 40–67.

World Bank Group International Finance Corporation, Environmental, Health, and Safety Guidelines, *General EHS Guidelines: Environmental Air Emissions and Ambient Air Quality* (April 2007).

World Bank, *China: Air Land and Water, Environmental Priorities for the New Millennium* (2001).

World Bank, *Program Document for a Proposed Environmental Sustainability Development Policy Loan* (Sept. 5, 2008).

World Bank, *Project Performance Assessment Report, India Environmental Management Capacity Building Technical Assistance Project 28* [Report No. 44250] (June 23, 2008).

Zaneski, Cyril T. and Margaret Kriz, "Deja Vu All Over Again: Another Bush, Another Clean Air Debate." *National Journal/Con gress Daily* (2001): 41–42.

Zorn, Graham, "Prevention of Significant Deterioration and Its Routine Maintenance Exception: The Definition Of Routine, Past, Present, And Future," 33 *Vermont Law Review* 783 (2009).

[illegible] Stabilization [illegible] and Coastal Ecosystems [illegible]

[illegible]

[illegible] Development [illegible]

[illegible]

[illegible]

Printed in the United States
By Bookmasters